AF666973

The Futurica Trilogy
Part 1

Alexander Bard
Jan Söderqvist

ALEXANDER BARD JAN SÖDERQVIST

THE NETOCRATS

The Futurica Trilogy, Part 1

Futurica Triology
The Netocrats

Alexander Bard & Jan Söderqvist © 2024
Aniara Press AB, Stockholm 2024
Photography by Pierre Björk
Translation by Aniara
www.aniara.one
Info@aniara.one

ISBN: 978-91-90003-25-1

Book cover & typesetting by
Per Gustafsson (modernstyle.se)

TABLE OF CONTENTS

1.

Technology as the driving force of history

There is a popular story that tells how a Japanese soldier was found several decades after the end of the war in an inaccessible part of the Asian jungle, where he had single-handedly carried on fighting the Second World War. As a result of a combination of circumstances he had been left there alone. Perhaps he had been ordered to remain at his isolated post, and had been exercising his duties to the fatherland with exemplary loyalty for all those years, or perhaps he had simply been too frightened to venture into populated areas. But time had passed and no-one had told him that peace had been declared. So the Second World War was still raging inside his head.

We have no reason to laugh at this confused soldier. He may have been wrong, but then so have we, countless times. The soldier was not particularly well-informed, but then neither are we always. We all suffer to some extent from confused perceptions of what is going on outside the small part of our immediate world that we can get a direct impression of. This does not prevent us from forming, and being forced to form, opinions about one thing after the other, even in complicated matters where our knowledge is limited to say the least. Most of what we believe that we know is precisely that: what we believe ourselves to know. Other people's actions are comprehensible to us only in so far as we actually know what they in turn believe themselves to know. Which is something we seldom know. The constant inadequacy of this information means that we have to swim through an ocean of misunderstanding on a daily basis, an activity which is both demanding and costly.

Like the Japanese soldier we form our lives inside our heads. We have to, because the world is far too large and complicated for us to open ourselves to its every aspect without protecting ourselves with a multi-layered mental filter. For this reason we create fictions for ourselves, simplified models of how we believe the world works, or how we think it ought to work. These fictions have to fill the immense vacuums between our limited areas of knowledge. It is within this world of private fictions that we think and feel, but it is outside in the collective reality that our actions have their consequences. The more complicated a situation, the higher the degree of guesswork and the greater the contribution of fiction to our perception of reality.

This dependence upon fictions often has dramatic consequences, not just for us personally but for society as a whole. Like the Japanese soldier we are fumbling blindly through dark forests. We react to signals that we can only partially understand, the consequences of which are only partially visible to us. Important political decisions are based upon shaky foundations and often have completely different results than were foreseen; great weight is placed on diffuse expressions of opinion, most often in the form of general elections, which are in turn the result of minimal knowledge, a problem which has been discussed, amongst others, by the author and journalist Walter Lippman in a couple of perceptive and intelligent books. This increasing lack of an overview explains for instance why the today's voters find it easier to understand the credit card fiascos and alcohol consumption of individual politicians than serious political issues. Symbolism becomes attractive when real problems are perceived as being far too complicated. The business world is constantly forced to redefine its prognoses and adjust its decisions retrospectively in order to conceal the fact that they were based upon fictional rather than factual conceptions, as a result of the perpetual and chronic lack of information.

Becoming informed is an attempt to synchronise your own head with the reality outside. There are good reasons to make the effort: it is easier to interact with your surroundings when you have a relatively correct understanding of its mechanisms. Someone who has educated themselves in the psychology of the stock market has better prospects of succeeding in the markets; someone who has educated themselves in their own and other people's inner needs has better prospects of succeeding in relationships, and so on. Every failure reveals that we were not as well-informed as we thought or had hoped. The discrepancy between our own and other people's perception of reality, and between our own fictions and actual reality, was far too great. We learn from our mistakes; we take account of our earlier failures in the future and adjust our behaviour accordingly. To put it another way, we make use of information.

Fictions can be more or less truthful, more or less applicable. They come in all possible forms, from private hallucinations to scientific theories. We are constantly testing them. Our culture consists of a perpetual evaluation and combination of both seemingly promising fictions and already proven fictions. The relationship between the fictions in our heads and unaccomodating realities is a recurrent theme in literature. Don Quixote, Othello, Raskolnikov and Emma Bovary are all victims of their own feverish ignorance. They are all relatives of the Japanese soldier. In attempting to study and gain an impression of the world around us we have to learn to differentiate between our prejudices – simplified models that we make use of not because they reflect empirical evidence but because they appeal to our own personal interests – and factual analyses and prognoses – necessary and intelligent simplified models of reality which make it comprehensible to us, even if the results do not appeal to us or fit in with our cherished fictions.

Our thoughts are directed by access to information. The story of the Japanese soldier is an illustration of this: without access to news from the outside world he lived out an imaginary war for several decades.

The same thing applies to whole societies and civilisations. Available information dictates which thoughts and actions are possible. It was not a lack of raw materials that prevented the Vikings from using water-skis or the Romans from videotaping their orgies – it was a lack of relevant information. Civilisation, in essence, is a matter of information. This means that any technological development which dramatically alters the preconditions for actions and the dissemination of information also implies a thorough re-evaluation of old and ingrained patterns of thought. The consequences of such a technological revolution are defined as a new historical paradigm.

The advent of language was one such revolution. The apes, our closest relatives, are intelligent animals with fantastic learning capabilities. But we cannot teach them to speak. From a physiological perspective we can say that their upper airways cannot function as vocal organs. But apes cannot use sign-language in any real sense either. Chimpanzees can learn to combine signs in order to communicate on the level of a small child; they can indicate that they want something or that they want someone else to do something, but they never exchange experiences, never speculate about the great mysteries of life. They lack the capacity to communicate their thoughts and experiences with linguistic symbols, which seriously hampers the exchange of information. Man's path diverged from that of the apes about five million tears ago, but language took longer to develop. To begin with we had elementary problems with our vocal organs, and evolution is a slow process. It is difficult to specify an exact time for the advent of spoken language, but current research suggests that it occurred as recently as 150,000-200,000 years ago. Only when the development of both the brain and our anatomy was sufficiently advanced was spoken language possible.

Language differentiates man from other animals. The creation of technology requires abstract thought, which in turn arises from a linguistic system of symbols. Language made it possible for man to develop socially and to gather and maintain collectives, which opened up a new world of interwoven relationships between individuals. Social life developed entirely new and rich nuances as communication became more advanced. Language offered the possibility of innovative thought, with all its countless possibilities of expression, and stimulated creativity and intelligence. It also made possible the dissemination of information to everyone who was connected to a community. The basic facts of life for a hunter-gatherer society – which plants are edible, which poisonous plants are edible after various treatments, which animals leave which tracks, and so on – became possible to communicate throughout a large group, and between generations. Other people could gain knowledge of both successes and failures, and could go on to develop further the combined experience of the collective. Mankind developed memory. Knowledge could develop, but only to a certain point. Spoken language does not permit, at least not without a tape-recorder, the reliable and comprehensive storage of information.

The mathematician Douglas S. Robertson has calculated the combined amount of information that a group or tribe of linguistically capable but illiterate people can access. He takes the poem The Iliad as his basis, a work comprising approximately five million bits (one bit indicates a choice between two alternatives: yes or no, black or white, one or zero), and which we know it is possible for one person to memorise. If the amount of information that a human brain can store is h, then h would appear to be somewhere between one and two Iliads, or, in other words, somewhere between five and ten million bits. If we multiply h by the size of a prehistoric tribe, a number between 50 and 1000, we get the maximum amount of information available within a society that was not capable of writing. We ought to bear in mind that there is a sizeable amount of redundant information here. Large amounts of the total store of information – how to hunt, how to fish, and so on – can reasonably be assumed to have been shared by most members of the community, which means that the total amount of information must be adjusted downwards accordingly. The numbers themselves must, of course, be taken with a pinch of salt, but Robertson's calculations provide an excellent illustration of the impact of written language when it was developed during the third millennium BC, and of the explosion in the amount of available information this represented.

Four of the so-called cradles of civilisation – Egypt, Mesopotamia, the Indus Valley and China – developed at roughly the same time, and what united them, and simultaneously differentiated them from the surrounding societies in which trade and metallurgy were also practised, was the invention of written language. To begin with clay tablets were used to write on. The earliest 'book' consisted of several of these tablets, stored in a leather bag or case. Certain texts, laws for instance, where inscribed on large surfaces so that everyone could see them. In this way the fundamental ideas and norms of the society were transformed from something mystical and ancient which had been communicated orally by shamans, into a visible and limited number of clauses and decrees that were available to everyone. Primitive, closed societies assumed a more open and more complex character. At the same time it became clear that knowledge gave power. Early forms of writing were initially an instrument of power. The Sumerian kings and priests used scribes to work out how many sheep different people ought to pay in tax. Another use of writing was propaganda: the ruler reminded his people of who was in charge and of the glittering victories he had won for them.

It was never intended that the written word would come into the hands of every Tom, Dick and Harry. The purpose of the first writings was, in the words of the French anthropologist Claude Lévi-Strauss, 'to facilitate the enslavement of other people'. But revolutions have their own velocity, impossible to control for any length of time, and this is particularly true of information technology. Things that occurred either long ago or far away assumed a completely different accessibility and visibility when communicated via written text. The amount of available information exploded thanks to the ingenious invention of a visual code for communication. Intellectual life became far more vital. Thanks to the phonetic alphabet – where each sign represents a sound instead of a word or concept – the ancient Greeks were able to develop philosophy and sciences that had a far more firm structure, a grammar. The replacement of the ear by the eye as the main sense of linguistic reception brought with it a radical change in mankind's way of understanding the world.

Written language looked like magic: it was entirely logical that the Egyptian god Thoth, who gave the gift of writing to mankind, was also the god of magic. Reading and writing transformed both knowledge and the world. Empires could be established and held together only when written communication had developed; only then was it possible for detailed information such as orders to be communicated across large distances. This led to the dissolution of city states. The decline in papyrus production during the reign of the last emperors is held up by many historians as one important reason for the decline and ultimate collapse of the Roman Empire. Even hand-written information had its limits.

Johann Gutenberg's invention of the printing press in the middle of the fifteenth century was the start of the next epoch-making revolution in information management. The printing press was also a basic precondition of what became modern science, and of the great discoveries and technical advances that led to industrialisation. Printed books were the source material of the astronomer Nicholas Copernicus, and without the printing process his manuscript may well have gathered dust on the shelves of a monastery library. Instead his De Revolutionibus, the thesis proposing for the first time that the Earth moved in orbit around the sun, spread quickly across the world of learning, where nothing was ever the same again.

Once the ball had started rolling, nothing could stop it. To put it bluntly, the printing press provided gifted and innovative people with the necessary information and inspiration to a previously undreamed of extent. Christopher Columbus read Marco Polo, large numbers of manuals and other technical literature circulated in Europe, and the whole of this tidal wave of new information prompted the development of new techniques and new thinking on the management of information, methods which paved the way for the gradual development of the sciences. Among the many innovations which followed in the wake of the printing press, after a certain incubation period, and which thoroughly and comprehensively altered mankind's way of looking at himself and the world, can be counted the clock, gunpowder, the compass and the telescope.

One illustrative example of the power of developed information management, provided by the physiologist Jared Diamond, is the historically decisive meeting between literate Europe and essentially illiterate America in 1532. In the city of Cajamarca in the Peruvian highlands Francisco Pizarro, with 168 men, captured the Inca leader Atahualpa, who had at his command more than 80,000 troops. The event only becomes comprehensible in light of the fact that the Inca leader knew nothing about his uninvited visitors whereas the Spaniards were well-informed about their opponent. Atahualpa was completely unaware that these visitors were in the process of conquering the whole of that part of the world, and that the great Indian civilisations of Central America had already fallen to them. He was entirely dependent upon defective oral information.

Atahualpa did not take the invaders seriously, and when his troops saw troops on horseback for the first time in their lives they panicked. Pizarro himself may not have been able to read, but he was a participant in a culture of writing and printing, and therefore had access to a wealth of detailed information about foreign civilisations. He was also aware of every phase of the Spanish conquest, and based his campaign upon the tactics of Hernando Cortés when he had defeated the Aztec leader Montezuma. Pizarro's success soon became known in Europe. In 1534 a book was published describing the events of Cajamarca, written by one of his company, which was translated into several other languages and became a bestseller. There was a great demand for information, and its benefits were self-evident.

Today's electronic and digital media comprise the most comprehensive information revolution of all. For a long time we believed that the central purpose of the computer was to think, to produce an artificial intelligence that would far exceed our own. Many people claimed that this goal was within sight when a computer named Big Blue beat the world master Garry Kasparov at chess. Today we can see that technology was heading in a different direction, towards communication via networks. Increasingly powerful and fast computers are making possible infinitely complicated and time-consuming calculations and simulations which were previously impossible to perform, which is of incalculable benefit to mathematicians and other researchers. Our collective knowledge is growing exponentially. But it is the global, digital network which is the most interesting aspect of this development. A new, dominant media technology means that a new world is evolving.

The Internet is something completely new: a medium in which virtually anyone, after a relatively small investment in technical equipment, and with a few simple actions, can become both a producer and consumer of text, images and sound. In this sense it is hard to think of anything more democratic; on the Net we are all authors, publishers and producers, our freedom of expression is as good as total, and our potential audience limitless. There are oceans of every conceivable sort of information available at the touch of a button. The growth of this new medium has been unparalleled.

The foundations of the Internet were laid as early as the 1960s with the decision of the American defence organisations to use computerised networks to decentralise their resources via a series of distant but connected terminals. The purpose of this was to protect against and limit the effects of any nuclear war with the Soviet Union. Eventually American and foreign universities were connected to the system after it had proved stunningly effective in the organisation of joint research projects. This development explains why the World Wide Web, the system which later became the standard for homepages on the Internet, was developed not in the USA but by researchers at CERN, the European institute for research into particle physics in Switzerland.

It was not until the end on the 1980s, as a direct result of the breakthrough of the personal computer and the launch of telecommunications modems, that the Internet was transformed from a tool for the military and the scientific communities into public property. Even in the early 1990s there were relatively few people who had heard of the Internet. It was only in December 1995 that Bill Gates woke up and announced that Microsoft would be changing direction and concentrating on Net traffic. Since then the growth of the Internet has been phenomenal. It is practically meaningless to give any figures regarding the number of computers linked to the Net because its development is so dizzyingly fast. Figures that were accurate when this was written will be hopelessly out of date by the time it is read.

There are various responses to this development. Critics suggest that all this talk of IT-revolutions and new economies is preposterous, or at the very least seriously exaggerated. These sceptics often point to the fact that even if IT- related shares are soaring on trend-sensitive stock markets the world over, most of these companies are posting continual losses, and that this cannot continue in the long run. The only people who have become rich from computers and IT are the various consultants and the producers of the computers and the software that make the Internet possible, while consumers have invested heavily for little or no gain. Any reflected exponential growth in the economy as a whole has not materialised.

From the point of view of the sceptic, the world is essentially the same as it was. We still manufacture and sell hammers and nails, the banks continue to devote themselves to the lending and borrowing of money, a few office routines have changed, but the significance of all of this has been exaggerated. Most people now write their own business letters on a word-processor instead of using a dictaphone or a secretary, but the question is whether the state of things has been dramatically improved by this. What is known as e-commerce is just business as usual, even if we are using flashy new machines. According to this point of view, this is largely a case of following trends, that there is a certain cachet in being first with the latest innovations, no matter what concrete benefit these may actually bring. And it matters little what technology we use to communicate: it is still the content which is important. Old and tested truths will still be just that in the future.

The contrary point of view is ecstatic. Anyone who has seen the light on their screen claims that everything will automatically turn out for the best. The Internet is the solution to all our problems: the economy will blossom for everyone forever, ethnic and cultural conflicts will fade away and be replaced by a global, digital brotherhood. All the information that becomes available will make our duties as citizens more meaningful than ever, and the whole of the democratic system will be revitalised as a result. In the digital networks we shall find the social cohesion that we often lack today, and harmony will spread throughout society. Entertainment will become, thanks to the inexhaustible possibilities of this new technology, more interactive and hence more entertaining than ever.

Both the sceptic and the enthusiast are mistaken. Neither radical scepticism nor blind faith is a fruitful strategy for orientation in the accelerated process of change in which we find ourselves. Both of these points of view indicate in essence an unwillingness to think critically, an inability to see. They are not analyses or prognoses, but prejudices. A new, revolutionary technology for communication and information will undoubtedly change the preconditions of everything: society, economy, culture. But it will not solve all our problems. It would be naïve to believe that it could. Development means that we can approach certain problems in a dramatic way, but to balance this we will have to confront a whole raft of new problems. We can live longer and more healthily, perceive ourselves to be freer, and realise more of our dreams. But the fundamental conflicts between classes and groups of people are not going to go away, just develop into more intricate and impenetrable patterns and structures.

Change of this type is not instantaneous. The sceptic who triumphantly points out that most of the global economy is still based upon the production of physical objects like fridges, aeroplanes and garden furniture rather than digital services on the Net is partly a little impatient – we are still in many respects only in a preliminary phase – and partly incapable of grasping the extent of the change. There is no question of the fridge disappearing, but rather that the objects around us will take on new significance and new functions in an entirely new socio-ecological system. Marketing campaigns for fridges, for example, will no longer stress their capacity to keep milk cold, because we take that for granted, but rather their capacity to communicate intelligently in a network.

It is in the nature of things that it takes a certain amount of time for changes to be absorbed. Every revolutionary technology only reveals its true colours after an unavoidable period of incubation. As far as the printing press was concerned, it took more than three hundred years before it made its definitive breakthrough, the point at which it caused a dramatic shake-up of social structures and created a new paradigm: capitalism. It took time, quite simply, before literacy was sufficiently widespread for print to affect large social groups beneficially. It was not until the Enlightenment of the 1700s that thinking became sufficiently modern, the exchange of information sufficiently lively, and technical advances sufficiently explosive for there to be signs of nascent industrialism in the offing.

Literacy spread rapidly through northern Europe during the 1600s, but its growth only accelerated more noticeably during the following century, primarily as a result of Protestantism and the dissemination of Bible translations into the various national languages. The preconditions were created for a completely new sort of critical public life, whose platform was primarily the first newspapers of recognisably modern form. New publications, such as The Spectator in England, were aimed at (and therefore also helped to shape) an educated and cosmopolitan middle-class. The aim of the newspapers was to inform about and debate the latest ideas. In France the world of the salon arose, where the aristocracy and middle-classes came into contact with one another and together examined the signs of the times. This form of gathering quickly became popular and spread throughout Europe.

But even if literacy and the development of information technologies lay the basis for the changes that occurred in society, they cannot explain them fully. A whole mass of factors have to coincide and co-operate if any epoch-changing process of change is to be set in train. The French sociologist Jacques Ellul, whose interest is primarily with the internal logic of technology and its radical effects upon our lives and environment, has pointed out a number of key phenomena. The first and possibly most self-evident precondition is that the necessary apparatus must be in place already, which in turn presupposes a longer historical process. Every innovation has its roots in a previous era. Novelty consists of what can be termed a technical complex; in other words, a series of inventions of various sorts which together form a powerful combination which is stronger than their individual parts. Innumerable innovations saw the light of day between 1000 and 1750, many of them remarkable in themselves, but they played to different tunes, they did not communicate with one another. It was only after 1750 that innovations began to work together and thereby facilitate large-scale industrialisation.

Another important precondition, according to Ellul, is population growth. An increase in population means increased demands which cannot be satisfied without growth. Necessity is the mother of invention. From another, even more crass, point of view, an increase in population means greater preconditions for research and technical and economic development partly in the form of an increase in the size of the market, and partly by providing a human basis for various experiments with different types of product. A third effect is that two specific and at least partially contradictory demands are placed upon the economic environment, which has to be both stable but also in some form of dissolution. On the one hand, a stable base is required for scientific experimentation which is necessary but unprofitable in the short term, but on the other there must be a capacity for widespread and fast change, a willingness to stimulate and absorb new thought-processes. The fourth precondition concerns the social climate itself, and is, according to Ellul, probably the most important of them all. There has to be a loosening of various religious or ideological taboos, and liberation from any form of social determinism. For the development of industrialism, for instance, it was vitally important that a whole raft of traditional ideas about what was 'natural' were thoroughly revised. No longer were either nature itself or hierarchical social orders perceived as sacred and inviolable.

Perceptions of man and his place in the world underwent radical change. The individual gained a new position, and human freedoms and rights were spoken about, which undermined preconceptions of natural groupings and classes. Suddenly unimagined opportunities opened up, offering social advancement and an improvement in living standards. The liberation of the individual and increases in technological efficiency co-operated. An historical resonance arose, where various factors dramatically strengthened one another in an accelerating spiral. The middle-classes were rewarded for their willingness to adapt and made the most of this opportunity. Hence the middle-class became the dominant class of the paradigm of capitalism.

The Industrial Revolution meant that that mankind's physical power was multiplied may times over through the use of machines. The Digital Revolution means that the human brain will be expanded to an incomprehensible degree through its integration with electronic networks. But we are not there yet, the necessary preconditions are not yet in place. Technology may be accelerating with breathtaking speed, but we humans are slow. Once again we are hampered by all kinds of religious and ideological taboos. Once again we are on the brink of a period of necessary creative destruction. This development cannot be controlled to any great extent. History shows that every new technology worth the name has, for better or worse, 'done its own thing', completely independently of what its originators had imagined. In the words of the communications expert Neil Postman, technology 'plays its own hand'.

Take the clock, for example, an apparently neutral and innocent artefact, but actually an infernal little machine that creates seconds and minutes, which has retrospectively given a whole new meaning to our perception of time. When the first prototypes were developed by Benedictine monks during the 12th and 13th centuries their purpose was to establish a certain stability and regularity to the routines of the monastery, principally with regard to the prescribed seven hours of prayer each day. The mechanical clock brought precision to piety. But the clock was not satisfied with this. It soon spread beyond the walls of the monasteries. It may well have kept order over the monks' prayers, but above all else the clock became an instrument which synchronised and watched over the daily lives of ordinary people. It was thanks to the clock that it became possible to imagine something like regular production during a regulated working day. It became, in other words, one of the cornerstones of capitalism. This invention, dedicated to God, 'did its own thing' and became one of Mammon's most faithful servants.

The same thing happened to the printing press. The devout Catholic Gutenberg could scarcely have imagined that his invention would be used to deliver a fatal blow to the authority of the Papacy and promote Protestant heresies by making the word of God accessible to everyone, which in turn made everyone his own interpreter of the Bible. When information became generally available, the natural but no less unforeseen consequence was that various accepted 'truths' were put into question. From the 1700s, modern rationalism developed alongside the notion of the educated citizen, and it was the printed word that was to do the job. The goal was the extinction of every form of superstition, principal amongst them religion and the monarchy. According to the French Enlightenment thinker, Denis Diderot, 'Man will not be free before the last king has been strangled with the entrails of the last priest'.

As long as information was an exclusive rarity, confined to the privileged few, it was unthinkable that ideas like that could be widely disseminated. Instead it became, after an incubation period of two hundred years, a mass movement. Technology played out its hand. And in the process, everything was changed. When the true agenda of the printing press began to appear, there was no longer any question of the old Europe plus a nice new invention, but of a completely new Europe which thought and acted in new ways. The progression had been uncovered, the historical process began to become clearer, and common-sense and science would lift mankind out of the darkness of ignorance and progressively improve standards of living. A new world view, and a new view of man, had been born.

A new, dominant information technology changes everything, not least language. This is partly because of new terminology, new words for new toys, but the most interesting and, to an extent, most problematic aspect of this is that old words assume new meanings. As language changes, so does our thinking. New technology redefines basic concepts such as knowledge and truth; it re-programmes society's perceptions of what is important and unimportant, what is possible and impossible, and, above all else, what is real. Reality assumes new expressions. This is what Neil Postman means when he talks of society going through an 'ecological' change. Technology shakes up the kaleidoscope of our intellectual environment and world of ideas and shows new, unforeseen patterns. We are entering a new social, cultural and economic paradigm.

The paradigm defines which thoughts can be thought, quite literally. The paradigm is simply the set of preconceptions and values which unite the members of a specific society. To take one example: when 'everyone' at a certain point in time is convinced that the world is flat, it is pointless to try to work out a way of sailing round the world. When Copernicus claimed that the Earth actually moved around the sun many people thought him mad. This is no surprise. Ridiculing his critics with the benefit of hindsight merely proves that one does not understand how a paradigm works. It is not possible to say categorically that his critics were wrong, because what they meant by the term 'Earth' was precisely a fixed point in space.

The terms still carried their former meanings, the paradigm shift had not yet taken place, people were still thinking along ingrained lines. The same thing occurred with the transition from Newton's physics to Einstein's. Many people dismissed Einstein's general theory of relativity for the simple reason that it presupposed that the concept 'space' stood for something which could be 'bent', when the old paradigm dictated that space was constant and homogenous. This was wholly necessary – if space had not possessed just these qualities, Newtonian physics could not have functioned. And since Newtonian physics had apparently functioned well for such a long time, they could not be abandoned easily. Hence a situation arose in which two paradigms competed with one another.

But two paradigms cannot exist for one person at the same time. It is either/or. The Earth cannot be both mobile and immobile at the same time, space cannot simultaneously be both flat and curved. For this reason individual transitions from one paradigm to another must be instantaneous and complete. It is like the Japanese soldier leaving the jungle and suddenly realising that he has been living an illusion for years: peace, not war, is the status quo, and Japan has become the driving force of the Asian economic miracle. We are speaking here qualitatively rather than quantitatively. To move from an old paradigm to a new is not merely a question of becoming informed in the sense of adding new facts to old ones with which we are already familiar, but rather in the sense that new facts, and old facts in a new light, change our world view entirely. And once we have perceived that our old world view is exactly that, old, and is no longer capable of explaining difficult phenomena, which it is in turn no longer possible to ignore or deny, then it is necessary to abandon large amounts of irrelevant knowledge. This is one of the sacrifices demanded by a paradigm shift.

From a narrower perspective this is an acute situation for someone trying to orientate themselves in the world which is being formed around us within and by the electronic networks. The problem is no longer a lack of information, but an incalculable excess of it. What appears to be new information and new ideas might actually be yesterday's news, or in the worst cases abject nonsense, which will direct us into time- and resource-wasting cul-de-sacs. Old recipes for success become outdated fast. It is only human to become more attached to old strategies if they have proved successful in the past, and it is therefore all the more difficult to abandon them. Someone who has built up a successful business, or who has merely managed to make his life tolerably comfortable, seldom recognises the necessity of dropping everything and starting again from scratch.

It is here that we find the true novelty in what is happening now. Previously the point of a paradigm was that it provided us with firm ground beneath our feet after a longer or shorter period of tremors. We need to get accustomed to losing that luxury and recognise that change itself is the only thing that is permanent. Everything is fluid. The social and economic stability that has been the ideal and the norm is becoming more and more the exception and a sign of stagnation. It is not enough to think, or to think in new ways; it is now necessary to rethink constantly, and to think away old thoughts. Creative destruction never rests.

Within the world of scientific theories, where the concept of paradigms was first established, there is talk of anomalies and crises. Anomalies are phenomena which are in part unforeseen, and in part difficult to adapt to fit the current paradigm. We can see them all around us these days: in society, within our cultural life and media, and in the economy. The preconditions which underlie politics are altering at a dizzying pace. Yesterday's ideological maps have nothing to do with the reality of today. Whole branches and great empires within the media are collapsing before our eyes. Working life is undergoing a dramatic revolutionary process which is effectively destroying all our old preconceptions of secure employment, automatic promotion and hierarchical organisation. Youngsters still wet behind the ears and wearing strange clothes are becoming multi-millionaires in a few short months, in businesses which few of their share-holders have any real grasp of.

When a large number of anomalies appear there are two possibilities. The first is to try to squeeze the new phenomena into the old system of explanations. This is what people have always done within science: patched up and repaired old theories, like for example the old Ptolemaic system of astronomy with the Earth in the centre and all the other heavenly bodies circling around it. It holds for a while, bearably, but with time it becomes gradually more apparent that the conditions produced by the old theory are no longer of any use. And then we are confronted unavoidably with option number two: to admit that the old system has had its day, even if there is no new system ready to take its place. This precipitates a crisis. The importance of this crisis is that it signals a need for new thinking. And this is where we are at the moment, in the middle of the crisis which has arisen from the old capitalist paradigm showing that it is incapable, but before any new system has won over enough adherents to be able to function as a generally accepted explanatory model. A lot of people are still patching up and repairing the old system, and there is a noticeable lack of new thinking. Sullen scepticism as to whether the new is actually anything genuinely new, and blind faith in the new which maintains that everything is now on its way to ordering itself automatically for the best, do not count as new thinking.

Writing about the future is obviously incredibly problematic because is does not yet exist. The best we can do is to produce more or less qualified guesswork. Someone who understands how dominant information technologies have played out their hands throughout history, and who understands how the dynamism within and between digital networks functions, has the best possible preconceptions for grasping the essential points of the current revolution. As an introductory note we claim two things. The first is that a new social, cultural and economic paradigm is taking shape. The main reason is the ongoing revolution within the management of information: digitalisation, and the astonishingly fast development of electronic networks. One immediate consequence of this is that our mental ecology is drastically changing, which in turn forces a whole sequence of necessary adjustments. And secondly we suggest that the form that the new paradigm is in the process of assuming will not be concrete, but fluid. It is not merely that we are developing new social norms; it is a matter of a completely new sort of norm.

The Japanese soldier in the jungle was ill-informed, and was fighting his own private world war within his own head, but then his circumstances were hardly optimal. We, on the other hand, cannot blame anything other than laziness or stupidity if we do not manage to garner a relatively clear picture of what is going on around us and if we cannot draw the relevant conclusions from this picture. Because one thing we can say without any doubt is that it will not be the meek who inherit the earth.

2.
FEUDALISM, CAPITALISM AND INFORMATIONALISM

Of course, the meek have never inherited much of anything. In every age, the power and the glory have belonged to those who are receptive and industrious, those who are sensitive to changes in the prevailing climate, who look out for their own interests and those of their particular group, and who have the great good fortune to be blessed by historical developments moving in their particular direction. Every new paradigm creates its own winners and losers. Altered preconditions for development and adjusted measurements of social status benefit new, up-and-coming social classes to the detriment of others. In order to understand the consequences of the current shift from capitalism to informationalism it would be helpful to take a look at the most recent historical paradigm shift, from feudalism to capitalism, and then compare the realignment of power distribution that occurred then with what is happening to us now. As we shall discover, there are so many parallels, on so many different levels, that we have little choice but to define what is happening in our time as a genuine paradigm shift of similar magnitude.

We can use cartography as a way of illustrating these recurrent patterns in historical developments. The most applicable concept in this instance is the mobilistic diagram. With inspiration from the work of philosophers like the nineteenth century's Friedrich Nietzsche and

Charles Darwin and the late twentieth century's power-theoreticians Gilles Deleuze and Michel Foucault, we can take as our starting point the idea that existence is a continual conflict between a multitude of different forces that are in permanent opposition to one another, yet which presuppose and define each other precisely as a result of this opposition. The main point here is not the forces themselves, but the tension in their relations to each other, how positions of strength are maintained or displaced, and the eternal struggle between them. Interaction, confrontation and communication are key terms here. By illustrating these complex relationships we can make them easier to identify and understand.

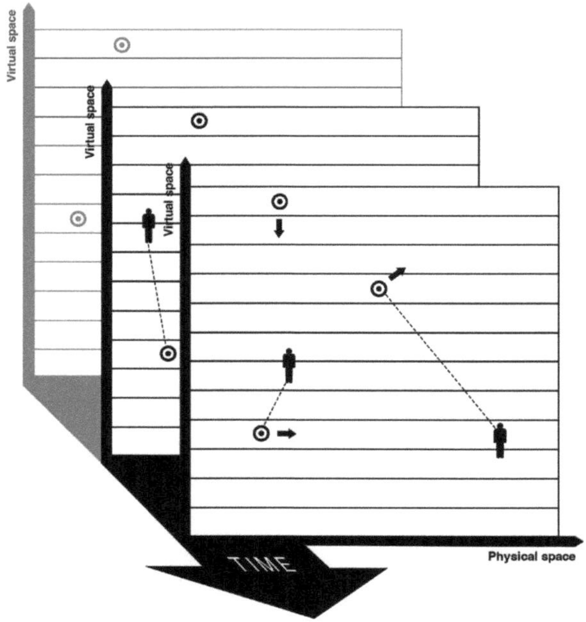

fig.2.1 Diagram of mobilistic power

Within the mobilistic diagram we can complement the Nietzschean-Darwinian two-dimensional plan of the existential conflict with a third dimension, a temporal axis which makes it possible to identify at any given point in time the specific value around which the conflict revolves. In other words, we are talking about an assumed point which we can allow ourselves to identify precisely because we are aware that we, the observers, are in a position of constant flux ourselves (the fact being that we in our role as viewers are just as much a force in this as those on the two-dimensional plan). This conflict is about power. The closer the interaction between any two complementary and/or opposing forces gets to the assumed point in this extra dimension of the mobilistic plan, the more power is at stake.

It is important to remember that this assumed point is not the only value that is the subject of conflict, but rather the central value in society, the defining feature of the paradigm in question. We might call this value "the religion of the time", or, more relevant to us, "the axiom of the time"; in other words, the basic suppositions in any given age about the structure of existence, the world view that is generally accepted and which is therefore socially functional. This assumed point makes it possible for us to orientate ourselves and focus upon what is going on. Because vested interests, not least the ruling classes of any given age, have expended a great deal of resources in the presentation of this assumed point as being not assumed, but real, in other words as being an eternal truth, then the fact that it is assumed becomes a highly sensitive area. When its assumed nature is revealed, the mat is pulled from under our feet, which can lead to a certain amount of giddiness. The common phrase "I don't know much, but there's one thing I know for certain..." is a good illustration of this problem. We are prepared to acknowledge that our knowledge is limited, but we imagine that we need to know at least something for certain in order to orientate ourselves in life.

Within the mobilistic diagram power is a moveable phenomenon with no inherent value, a neutral concept. Power migrates, is captured and recaptured, in all directions. Identities arise purely in relation to other identities. Definitions must be constantly tested as circumstances change. What's what? Which force is which? Can they be kept apart despite the fact that they're drifting into one another? We are on the sidelines watching a feverish struggle for status; a struggle about who creates what, who can gain control over whom, who defines what and why, with the ever-present supplementary questions: at what price and at the cost of whom?

The relations between forces, their interaction, are the crux of the matter. The master cannot exist independently of the slave, just as little as the slave can exist independently of the master. Each is conditional upon the other. It is the slavery of the slave which makes the master a master, and both are engaged in the eternal struggle for recognition which, according to G W F Hegel, another of the nineteenth century's great philosophers, is the motor which drives the entire historical process. According to Hegel, it was the desire for recognition by other people that set off the struggle for prestige in the earliest social groupings which lay the foundations for later divisions of humanity into different classes. These struggles have continued to rage, keeping society in a constant state of flux, so long as different groups perceive that they have too little recognition and consequently believe that they deserve greater influence and higher status.

A paradigm shift occurs when the assumed constant is moved and undergoes a qualitative redefinition. It is no exaggeration to compare this shift in underlying values to an historical earthquake. Every other factor in the arena of conflict is fundamentally affected by the fact that its point of origin, the instance around which a society revolves, is suddenly in motion. The consequence of this is that the actors no longer believe that they know anything for certain. Everything is in flux. Some of the older actors remain frozen in their historical roles, marginalised at the point in the arena where the central value used to be. New forces, new actors step into the arena and immediately instigate a new struggle around the point towards which the assumed value is moving. When the shock has subsided the old actors have to find new, less impressive roles for themselves.

This sudden movement of the assumed point is countered at once by strong resistance from those who regard their position as being under threat. When people and social classes become conscious of the fact that the constant around which the whole of their lives has revolved, and which has formed the basis for their identity, is in motion, they generally react at first with strong denials – "this can't be happening!". After a while, when the changes can no longer be denied, their reaction becomes one of either resignation or aggressive opposition – "this mustn't happen!". This is strengthened by the fact that the old authorities upon which everyone has relied have a highly confused understanding of what is actually happening. A good example of this sort of process is the destructive conflicts experienced by the western world since the transition from feudalism to capitalism about the concept of God, and the inevitable death of God. For every new mental barrier in our world-view that scientists have demolished, for every new boundary that has been transgressed, God has been pushed one step further into the unknown by his large but gradually diminishing crowd of supporters. To begin with God lived above the firmament, then beyond the sun and the planets, then beyond the stars, before finally being despatched beyond time and space. But he had to survive at all costs. Axioms connected to decrepit paradigms are often remarkably tenacious of life, not least among marginalised groups and classes.

Many people simply have difficulty understanding that the concept of God arose in a different paradigm to their own, with a purpose specific to that era: to the advantage of certain then current interests at the expense of others. In feudal society God was, in mobilistic terms, an assumed constant whose existence was unquestionable. (Merely trying to scratch the surface of this constant was punishable by death). With the transition to capitalism the fixed structures which supported the concept of God collapsed. When the central value began to move, the foundations which had earlier seemed unshakeable suddenly collapsed.

God's majesty became relative, and it was now possible to question his very existence. Christianity fell into an abyss of doubt about its own legitimacy from which has never managed to recover. What we regard today as phantoms and demons once had a very real impact on people's realities. This is not a matter of theology, or of the weakness of evidence supporting the existence of God, but a matter of power. The authority of both monarchy and church rested ultimately upon the existence of something called God; God was the assumed constant and could therefore never be called into question under any circumstances. If doubts were allowed to spread, the whole power structure would have been in danger of imploding.

As a result of obstinate denials of any movement in the assumed constant, and unwillingness to surrender the claims of religion, atheism was elevated to the status of a new axiom, and became an oppositional and effective tool for the acquisition of power by the developing bourgeoisie. This is illustrated most clearly in the most spectacular social experiment of the capitalist paradigm: the communist project. Communism was an inverted form of Christianity, an expression of the age-old dream of heaven on earth that was entirely typical of its time. The communist faith was inspired by the idea of social improvement through human rather than divine agency. The new State was to be the tool; new Man – thoroughly rational and reasonable – was the utopian goal. This dream lay waste to entire nations and continents, and led to between 85 and 100 million people (for obvious reasons it is difficult to be precise) being slaughtered in peacetime for the greater good of the cause.

The existence of people who still defend the communist project is explained by the fact that this is a matter of religious faith, whose irrationality is a blind spot in the otherwise perpetual invocation of logic. The power of this faith was precisely a reflection of its original opposing force, organised religion, which refused to the last – in Russia, China and Latin America – to let go of power. It is quite conceivable that had the last Russian Tsar publicly declared his belief in atheism, he would have denied communism much of its appeal and thereby prevented the Russian Revolution. The demon of the assumed constant is so strong that even its antidote - and hence its equivalent in the following paradigm -inherits and exerts an almost magical influence on our thinking.

During the current transition from capitalism to informationalism we can see several parallels to the displacements that occurred during the last paradigm shift. What has been characteristic of capitalism, its blind spot, its assumed constant, has been the Human Project. To begin with, it is interesting to note that the Human Project in its most naked form, the Project of the Individual, has been elevated to the point where it is the last remaining life-raft for humanists and others with faith in humanity to cling to in a sea full of the wreckage left when the more glorious parts of the Project, such as communism, capsized one after the other.

This explains why all the political ideologies of capitalism during this final phase of the paradigm proudly proclaim in the introductions of their manifestos their fundamental faith in "the individual". Under external pressure capitalism has returned to its infancy and taken refuge in its philosophical origins, to pre-industrial thinkers like René Descartes and Francis Bacon. What we see are desperate attempts to re-establish the project, even as it is being relentlessly dismantled, in the form of hyper-individualism. By shouting loudly enough they imagine that they can breathe life into the corpse again. This ideological Frankenstein's monster goes by the name of libertarianism.

Just as the Protestant revivalist movements, which appeared in conjunction with the breakthrough of the Enlightenment, can be described as a supernova phenomenon, a sudden flaring-up of obsession with the old assumed constant, we are now experiencing similar supernovas in the paradigm shift from capitalism to informationalism. Today's hyper-egoism, hyper-capitalism and hyper-nationalism are all examples of this sort of supernova phenomenon. This development has arisen because the entire Human Project, the elevation of the individual alongside the state and capital, its pole star, and all its allied progeny – academic, artistic, scientific and commercial projects – have made up the fundamental axiom of the capitalist paradigm. They have been assumed to be eternal, a guarantee of stability, but they are now in motion. The great struggle has only just begun, and the death of the Human Project, like God's funeral, will carry on for a long time yet, and be accompanied by convulsive spasms. It is enough to remember the amount of interest and resources that are invested in this project to realise the degree of social trauma inherent in the developments which have just begun. This can't be happening, this mustn't happen! Nevertheless, collapse is inevitable, for the simple reason that this project is indissolubly linked with a paradigm that has passed its expiry date.

Of course it is difficult to try to localise the assumed constant of the new paradigm at this point, or to identify the forces that will struggle for power. A contemporary analysis from our position in the midst of the tornado of the paradigm shift can never be anything other than a construction of qualified guesses. As long as the assumed constant is in motion (and it will be for a long time yet) the variables are incalculably large, which makes the task similar to a meteorologist trying to forecast the weather several years in advance. This does not mean that any analysis is uninteresting or unnecessary. On the contrary. A critical examination of existing power structures is never more important than when a new class-society is developing. That is the only time when an observer can play an active role in the struggle surrounding the assumed constant. Analysis itself has a chance to become a constructive part of the historical process which is under analysis, and can become one of the many factors which influence the events under discussion.

Even before the assumed constant has settled upon a new fixed point, one force, the seeds of a new dominant class, begins to form around it. But is it really possible to talk of a new dominant class simply because the assumed constant has moved? Even if this constant has changed its character, why should this mean that the dominant force in society must change? Ought we not to assume that when the dominant class of the old paradigm realises that the constant around which it built its position is in the process of moving, it would seek out, ascertain and occupy this new position? The old dominant class would thereby become the new dominant class, albeit in a new guise. But there are several reasons why this is not the case.

To begin with, man is essentially a conservative creature. In situations like these there is a psychological term for a phenomenon known as cognitive dissonance which, put simply, states that we have a marked tendency to cling to old beliefs despite the fact that they are at odds with known facts. The reason for this is simply that the old beliefs are just that, old and familiar, and that we are therefore fond of them; they are part of what makes us mentally comfortable. This leads to intellectual sluggishness: we are prepared to make greater efforts to preserve the status quo in our heads than we are to learn new things. At the moment when we learn anything new we have to change our lives, albeit very slightly. For this reason our capacity to move across the historical map is in practice minimal.

One conclusion, therefore, that we can draw from analysing the mobilistic diagram is that our surroundings move and alter a good deal faster than we do. Our movement under these circumstances is primarily forced, a reluctant reaction to the maelstrom of social forces and information that make up our surroundings. It is a lack of satisfaction of our complicated and limitless desires, or to put it more correctly: the idea of this lack, the desire for desire, which compels us to consume. It is intolerance or narrow-mindedness in any given society that compels us to migrate. It is society, the system itself, which is constantly in motion, and individuals and groups are drawn against their will into the vortex and are forced to give up old and secure positions.

Because we are the sole observers of history, it is tempting to exaggerate our power over our surroundings and to see ourselves as the agents of free will, as the takers of initiative on the historical stage. But this is merely to give in to delusions of grandeur. Our scope for independent action is severely restricted. Those of our actions which are most visible on the historical map are generally easier to interpret as reactive rather than proactive. That people were attracted to communism and other grand utopian ideas is, amongst other things, the result of our constant need for adjustment to constant change. The lure of utopian dreams lies in their promise of rest, and there is a strong and widely-felt desire to put a stop to the movement that has been forced upon us. But if we are to put a stop to our own movement, then the historical process itself must stop, otherwise it will roll over us, and the historical process is by definition a process, something in motion. The end of the historical process would be the same as the end of society's, and therefore also our own demise.

This has been proven time after time throughout history. Every attempt to realise a utopia – communism is the most obvious example – and thereby put a stop to this constant motion, has led inexorably to the death of that society. Death is the only real alternative to constant turbulence. Buddha realised this over 2,500 years ago. We have to choose between nirvana, a state of permanent calm, or accept that everything around us is in constant motion and change, which brings with it an inescapable need for constant adaptation. The fact that our room for manoeuvre is so limited from a philosophical point of view makes us in practice slaves to the historical process. The Russian Tsar could not have been a committed atheist because he would thereby have been forced to recognise the illegitimacy of his position. He could scarcely deny the God upon whom his authority was based. And things went the way that they did.

During a paradigm shift change is so dramatic that the old dominant class is shown to be incapable of controlling the assumed constant which defines the new paradigm. A new dominant class develops at the point on the map where the advantaged group happens to find itself as a result of historical coincidence. The transition of a society from an old paradigm to a new is a protracted affair, which means that for a long while there is a considerable, albeit secondary and diminishing, residual value at the point where the old paradigm was focused. This acts as an incitement to the old dominant class to cling on to the old assumed constant. Even at the end there are doubters – this can't be happening, this mustn't happen! There is no need to learn anything new if there is any reason to avoid doing so.

It is therefore not the case that everything old becomes instantly worthless in a societal paradigm shift. Even if the central value, to take one example, moved from the ownership of land to capital in the transition from feudalism to capitalism, there was still an undeniable value in land-ownership. But the nature of its value changed. Land became a prized commodity. It is important to remember that it was the new dominant class that determined the value of land-ownership, which came to be expressed in monetary terms. The bourgeoisie bought and renovated old estates, transforming them into private playhouses and resorts for country pursuits, thereby signalling that the bourgeoisie were not only the lords of the burgeoning working classes, but also of the dominant class of the old paradigm, the aristocracy. It was the bourgeoisie who decided the new rules of the game.

Before this, country estates were not for sale. Their value lay in their heraldic shields and their proximity to the king's residence. Within the new paradigm these estates assumed a value according to new principles, the principles of the open market. They were given a price-tag. Their new value was decided by a whole range of new variables: the size and quality of their forests and farmland, as well as the new dominant class's desire to associate and play with the old dominant class, and to appropriate and display its traditional symbols. It did not take long before the old, traditional, feudal symbols of power lost their connection to power and were reduced to the state of faded and majestic curiosities whose value was purely nostalgic. The new dominant class had a right royal time (pardon the pun) with the old attributes and cast-offs of the aristocracy: the monarchy, the court, ancestral names and etiquette. The paradigm shift had completely stripped these of their metaphysical associations, and the bourgeoisie showed that everything had a price by buying and selling and marrying their way to the noblest titles. The aristocracy had little choice but to join in and swallow its humiliation; it was necessary to earn money, the overpowering value within the new paradigm.

The astonishingly crass business arrangements that were the result of the aristocracy's acute need of money and the bourgeoisie's desire for luxuries are a recurrent theme of nineteenth-century literature. The most cynical and entertaining observer of these transactions was Balzac, who was himself not above inserting a "de" before his surname to make it look more aristocratic. The old trappings of majesty were preserved, but their function was altered, and ceremonial costume became fancy dress. The same patterns are reflected today as the netocracy, the new dominant class of the information age, disrespectfully plays with the sacred cows of the bourgeoisie: individual identity, social responsibility, representative democracy, the legislative process, the banking system, the stock markets, and so on.

One of history's ironies is that the bourgeoisie's obsession with mass production – it was the printing press which instigated this stage of industrial history and which was of central importance in the capitalist revolution – undermined the market for heavily symbolic aristocratic treasures by flooding the market with cheap imitations. An artefact which had been unique was now merely the original, admittedly more valuable than all the copies, but whose aura nevertheless lost much of its attraction as a result of everyone surrounding themselves with exact replicas. Their value as status symbols was soon devalued.

Because it was the new dominant class which decided the rules of the game and decided the amounts on the price-tags of the old estates, the aristocracy became helplessly marginalised in the capitalist economy. As long as it had property to sell it could live on in style in the country, relatively less wealthy and, above all, ever more distanced from the centre of power and of society. Country estates were overshadowed by banks; family names and heraldry were replaced by financial empires and academic titles; the court and its jesters were replaced by parliament and political journalists. The stage was captured by new actors. Many of the new roles were similar to the old ones, but the script was newly written and the plot of the drama itself underwent drastic modernisation.

The assumed constants of the old and the new paradigm are so radically different from one another that any aspirant to a leading role in the new drama will need to learn an entirely different culture and a whole new set of values. The fact is that it is often easier for the old underclass to adapt to the cultural demands imposed by the dominant class of the new age than it is for the former dominant class. The stage is set for a major realignment. The old underclass has less to defend and lose, and finds it easier to learn new tricks and allow itself to be changed. In line with the thesis of continual historical movement we can say that acceleration comes easier to someone already in motion than it does to someone who is standing still. It often takes a while to realise that all the old recipes for success have lost their validity, and that realisation itself is difficult to handle – this can't happen, this mustn't happen! One example of this today is that it is often easier for the young immigrants to adapt to the new age's demand for cosmopolitan openness and cultural plurarchism than for their contemporaries from the homogenous and native bourgeoisie.

The members of the new dominant class have made no specific attempt to end up near the new assumed constant. They have simply had the good fortune to have been at the right place at the right time. Just as in nature, which is also in a state of constant flux, social evolution occurs arbitrarily: certain mutations turn out to be advantageous under current circumstances. It is not so much a matter of survival of the fittest as of survival of the best adapted. And what the best adaptation is changes with the circumstances. According to the principle of mankind's intellectual sluggishness, the new dominant class is made up of individuals and groups who by sheer coincidence happen to be close to the point where the new assumed constant ends up.

The bourgeoisie became the new dominant class in capitalist

society. And where were these capitalist entrepreneurs recruited if not from the cities where they happened to be? They were also brought up under a Protestantism distinguished by a strong work ethic. The bourgeoisie did not seek power, did not seize power – it landed in its lap. The bourgeoisie was given power. If we take a closer look at this new dominant class there is further evidence that those who are already in motion are favoured above those standing still. New recruits to the bourgeoisie generally came from the surplus of peasants, the lowest of the low in the old power structure, rather that the heirs to aristocratic titles and estates.

The sociological equivalents to biology's genes are known as memes, ideas or interconnected systems of ideas, and a comparison of the origins and spread of genes and memes shows similar patterns. Just as biology has Darwinian development, sociology has its own memetic Darwinism. By studying genetic Darwinism we can draw interesting parallels to show how memetic Darwinism functions. The history of biology is a tough, eternal struggle for survival and reproduction among a wealth of haphazardly occurring species in a constantly changing environment. Coincidence determines which species will flourish at the expense of others; external circumstances determine which are best suited to current conditions, and the others are sifted out. The various species compete for a limited supply of resources with similar species, and, moreover, with related varieties of the same species.

Nature never rests, which is why the criteria for which mutations are best suited for survival are constantly changing. Human interference in nature also influences the conditions of the eternal struggle for survival, benefiting some and harming others. One famous example is the moths which during the nineteenth century became considerably darker in colour in industrialised parts of England. As a result of the environment becoming more polluted, the darker moths were better able to escape predators as they sat on dirty tree-trunks and walls. The birch trees were no longer particularly white, so it was the darker moths who reproduced and spread most successfully, with the result that within a few generations the appearance of the entire species had altered.

The same level of coincidence affects memetic Darwinism in sociology. In the dense jungle of complex and often contradictory information that surrounds us, the memes which can most easily survive and reproduce under the prevailing circumstances are the ones which eventually appear to be the strongest, while the memes which cannot find a firm footing gradually fade away and come to be regarded as weak. But the difference between strength and weakness in this case is seldom visible in advance, at least not if you are staring yourself blind at the memes themselves and forget the environment of information technology and its development. The task of futurists is to map out the ecological system in which the various memes are fighting, and, taking that as their basis, to make a prognosis of the various memes' chances of survival.

The values and cultural baggage of every individual or group is made up of a number of memes. Which of these proves to be either "strong" or "weak" in the Darwinian sense in conjunction with a paradigm shift can only be determined in retrospect. In the same way that various genes have no influence on natural changes during a genetic Darwinian revolution, all memes are impotent in the face of the immense social powers which are set in motion by a paradigm shift. The carriers of both memes and genes can merely hope that they will be fortunate. As far as the basic, theoretical preconditions are concerned, there is little real difference between biological and sociological Darwinism.

In order to understand the memetic Darwinian process it is useful, once again, to use cartography as a tool, this time with human beings and memes rather than social forces as the variables. We can see existence as a three-dimensional space once again, with the present as a two-dimensional plane with two axes. In studying the variables of human beings and memes, the axes become physical and virtual space. The third dimension is time, which we can ignore for the time being. By freezing a moment in time, like a photograph, we get a two-dimensional diagram which makes it much easier to examine the internal relationships of any given society. We can choose to fix either people or memes on the diagram, which makes it possible for us to study the relations between them.

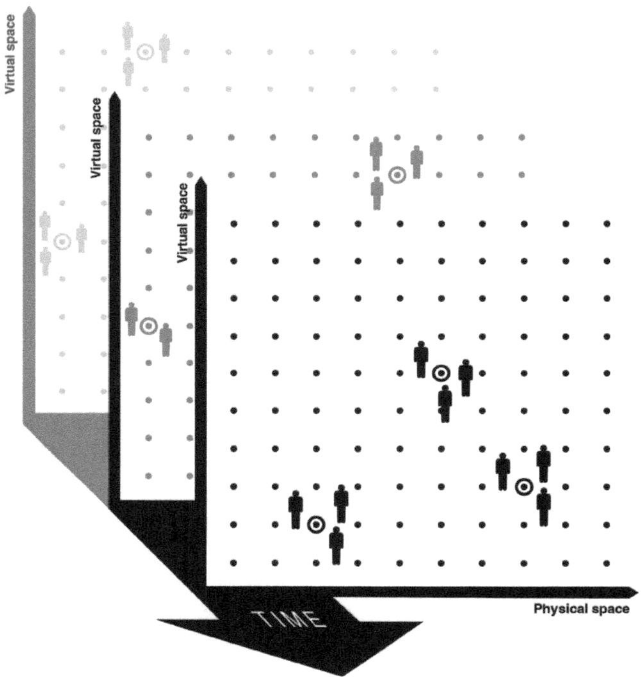

fig.2.2 Diagram of mobilistic identity

In our current example we can fix the memes, spreading them evenly across the diagram. What we discover when we study the concentrations of people is that these actors, the citizens of the society in question, are only attracted to a limited number of the available memes, and gather around these in noticeable clusters. They build their social identity on their relation to these clusters. The members of one cluster are "us", the members of the other clusters are "the others". It is important to remember that the actors do not choose their relations to the various clusters freely; their positions in respect to both the physical and virtual axis reflect a fact, not an ambition or aspiration.

At every frozen moment in time we can see that the largest cluster on the diagram is gathered around the meme which makes up the kernel of what we have called the assumed value of the governing paradigm. Under feudalism the court is one such central cluster, with the monarchy as its focal meme. Another strong feudal cluster is the church, grouped around the religious meme. Under capitalism trade is the most powerful cluster, with the banks and stock markets as focal memes. Other capitalist clusters of note are the apparatus of the state, grouped around the meme of representative democracy, and the academic sphere around the meme of science. In informational society the most important forum will be the Nexus, the portal of power, the linking node in the all-encompassing net. Gathered around this function will be the most important cluster of the informational paradigm: the netocratic network.

When we add the third dimension, time, we get a hologram. The first thing to notice is the rapid succession of memes: comprehensive production of them, and an almost similarly comprehensive destruction. Naturally it is the memes which are surrounded by the largest clusters which survive most easily. People vote with their feet. Out of all the religious memes that struggled for survival in Ancient Rome, only two are left today: Christianity and Judaism. The others fell victim to forgetfulness, the historical equivalent of creative destruction.

But the fact that certain memes prove to be attractive does not mean that they will remain unchanged through the centuries: on the contrary, they have to adapt and modify themselves continuously, to such an extent that we can speak of a ceaseless flow of new memes which have their basis in old ones. Most memes die and disappear, leaving space for new ones. At the same time, the memes which do survive have to adapt constantly, recreating themselves, in order to survive. The closer a new meme finds itself to an important cluster, or, in other words, the better suited a new meme happens to be to fulfil the needs and desires of the cluster, the stronger its chances of surviving in the ongoing struggle. Let us take one example: Bill Gates, a person who happens to be the world's richest man, was born in Seattle, a city which both physically, virtually and historically is situated close to the fast-growing technological industries of California. If Bill Gates had been born as a peasant woman in sixteenth-century Madagascar, we would obviously never have heard a peep about the Microsoft meme, which in turn would have altered the historical plane on which we find ourselves at the moment.

History shows time after time that people are far too conservative and sluggish to be able to move sufficiently quickly in any meaningful quantity to gain any advantage from the changes caused by a paradigm shift. Being aware of the fact that the assumed value is in motion, or that this motion will affect other important memes and clusters, is not enough to facilitate sufficiently rapid movement in the right direction. The fact that a Madagascan peasant is aware of events in Silicon Valley does not mean that he can go ahead and set up an internet company there. When it comes to the positions of individuals on the plan we are forced to recognise that coincidence – or fate, if you prefer – is decisive.

While capitalism was making its breakthrough, the aristocracy were busy with their country estates, far from the banks and marketplaces of the cities. Members of the aristocracy were bred to regard both trade and financial management with distaste. The old dominant class was fully occupied protecting its inherited rights to its family names and estates, in spite of the fact that the value of heraldic names in wider society was sinking rapidly. The aristocracy busied itself with polishing its heraldry and compiling glorious tales of the golden age that had passed. It had missed the boat, quite simply, abandoned by the passage of time. With the development of pietism, European Christians were encouraged to handle money and were allowed to charge interest on loans. Until then, this had been the preserve of the Jewish proto-bourgeoisie. The aristocracy did not stand a chance in the struggle for power in capitalist society compared to the bourgeoisie who happened to be at the right place at the right time, a vibrant mutation with its origins in the old peasant class, well-suited from a memetic Darwinian point of view to become the dominant class of the capitalist paradigm.

An interesting and noticeable phenomenon of every paradigm shift is the establishment of a secret pact, an unholy alliance, between the old and the new dominant classes. As soon as the transfer of power is an indisputable fact, there is a peaceful and discrete handover of power which is to the benefit of both parties. This secret pact is entered into with the intention of protecting both the common and distinct interests of both groups. The instigation of the pact is often followed by a time-consuming and noisy display of various pseudo-conflicts about meaningless symbols, all with the intention of hiding the existence and purpose of the pact.

The most important function of this secret pact is to secure the monopolies on public space of both dominant classes during the paradigm shift. It is in the interests of both parties to create the greatest possible confusion, the maximum amount of fuss, so that the transfer of power can take place as quietly and efficiently as possible, without the disruptive involvement of the underclass or internal critics. A classic example of such a pact between dominant classes was the eighteenth century's marriages between the sons of the aristocracy, with their inherited titles, and the daughters of the bourgeoisie, with their inherited capital. But this process needed to be complemented with a constructed pseudo-conflict to disguise the existence of the pact. It was of vital importance that neither party appeared to be part of a conspiracy.

This was the background to the still rumbling artificial pseudo-conflict surrounding the to-be-or-not-to-be dilemma of European and Asian monarchies. The aristocracy was permitted to retain its castrated royal families and its state-subsidised opera houses in exchange for its co-operation in managing and maintaining the capitalist nation state's various historically romanticised propaganda projects. It agreed to take on the role of "disarmed oppressor". The aristocracy was permitted to run museums and similar institutions where history was revised to make the existing social structure look "natural". When members of the aristocracy had sold all their family treasures and could no longer finance life on the family estate, and the daughters of the bourgeoisie had begun to prefer rich young men from their own class to titled poverty, they were permitted to remain on their estates under the condition that these were opened to the public at weekends. They were transformed into state-subsidised historical museums, slightly run-down, picturesque destinations for bourgeois family outings and Sunday walks. The aristocratic past was presented as a charming, tragic theatre-set, against which capitalist society and its bourgeoisie could bask in their self-proclaimed perfection.

By effectively fitting muzzles to both the aristocracy and the church, the members of the bourgeoisie could set about rewriting history, as if their own class and its nation state had always existed. The social constructions of the new paradigm were presented as eternal and "natural" truths. Man became God, science his gospel, the nation his paradise, and capital his holy instrument of power. This was the means of defence for the bourgeoisie's monopoly on power, history, language and even thought itself. The eternal truths of capitalism could not, would not and, indeed, never needed to be questioned. Behind this traffic in symbolism is hidden the important role of the secret pact in the development of the power structures of the new paradigm. As a result of the new dominant class arriving first at, or rather happening to find itself in proximity to the new central value, it was able to make maximum use of this advantage. This it did by accumulating vast wealth, generated by the assumed constant of the new paradigm, and all with the blessing of the old dominant class. The new dominant class achieved this coup de grâce by establishing a monopoly on public space, and then using this position first to deny the very existence of the new underclass, and later to deny its members any possible rights.

As soon as it became clear that land could be protected with the help of laws and a monopoly on power in the hands of the nobility, the fundamental basis for feudalism, the aristocracy took control of all available land. Not even the most remote woodland castle was left out, because it could then have been used as a future base for the peasants' demands for land-ownership. In much the same way the bourgeoisie, with the blessing of the aristocracy, were able to spend the first decades of industrialism plundering the countryside, and various colonies, of raw materials and labour, and running factories operated by slaves at an enormous profit. There is little reason to believe that the new dominant class of informationalism, the netocracy, will behave any differently to previous dominant classes. The increasingly marginalised bourgeoisie will come to be willing participants in this perpetually recurring historical drama, this time under the direction of the netocracy: a drama which has at its heart the denial of the existence of the new underclass.

In the same way that the aristocracy instigated the most important legal preconditions for the expansion of capitalism, state protection of private fortunes, the increasingly marginalised bourgeoisie will use its control of parliamentary legislation and the police to legitimise and protect the most important components in the construction of netocratic power: patents and copyrights. The principal precondition for the success of the new dominant class is therefore, ironically, a gift from the old dominant class. The morality of the new age is created around this handover of the historical baton. Just as the aristocracy and the bourgeoisie once enshrined the inviolability of private property in old laws, so the bourgeoisie and the netocracy are today united in their call for copyright as an essential defence of civilisation. Immense amounts of "science" are produced with the aim of proving its blessing for humanity as a whole. Within this strategy it is clear that any form of power which will not be helped by copyright is "immoral", which the legal monopoly of the bourgeoisie will immediately interpret as "criminal".

But, sooner or later, the secret pact of the dominant classes will be put to the test by the merciless demands of mobilistic analysis that any force can only be defined by an opposing force. We cannot talk about a new dominant class without at the same time defining a new underclass. The dominant class will use every available means to assert its right to total control over the assumed constant. But because this constant only attains its value by being recognised by the new class which is subordinate to the new dominant class, there is conflict about its value. The dominant class's relation to the assumed constant is based upon its desire to possess the constant, and to ensure its control over it. The underclass, on the other hand, is made up of those whose activities, in the form of production or consumption, or whose coincidental position on the historical map gives the assumed constant its value. The new dominant class's monopoly on public space ceases to exist when the new underclass becomes aware of itself, organises itself, makes demands and challenges the existing order. The master/slave relationship becomes tense and uncertain. Thus a new conflict commences, full of constantly recurring trials of strength, interrupted by temporary truces, only to explode into activity once again. It is from this conflict, this fight for power among the classes, that society and history gain their momentum.

When the aristocracy passed on the baton of real power to the bourgeoisie, there was a contemporaneous and ongoing formal transfer of power from absolute monarchy to directly elected parliament. There is never any comparable historical meeting between the underclasses of two different paradigms. The reason for this is partly that the old underclass is the main recruiting ground for the new dominant class, and partly that the two underclasses have no point of contact, which, ironically, is precisely because they are not in conflict with one another. Everything points to the same thing happening in conjunction with the breakthrough of informationalism. The new underclass, rendered practically invisible by the new dominant class, will long remain an unknown force, even to itself. In a society which is in every other respect overflowing with information, there is a telling lack of information on this subject. It is, once again, a matter of control over the production of ideology.

3.

Plurarchal society – the death of etatism and the crisis of democracy

According to the classical Marxist view of history, the rulers of any given society exert power over their subordinates by their control of the means of production. Power is the act of owning and directing the apparatus of production. The principal and overriding task of culture, seen from the Marxist perspective, is to justify the existing power structure by presenting it as "natural". The main occupation of feudal society was the production and distribution of agricultural products. Power was intrinsically linked to the control of land and its produce, and the dominant class of the feudal era, the aristocracy, was continually concerned with this: the control and legitimisation of its control of the land. Culture was put to good use keeping the underclass of peasants in its place. The existing order of things, the structure of social hierarchy and the aristocracy's limitless right to do whatever it wanted with the land were constantly presented as "natural" and eternally valid. This view of the world did not tolerate any alternatives.

The aristocracy's mandate to exercise unlimited power over the means of production was of divine origin, and was derived from a religion that was tailor-made for that purpose: defending the right to ownership of land. In feudal Europe, this part was played by Christianity. Stained glass windows in churches related improving stories in which obedience to one's masters was rewarded, while independence and/or self-interest were punished. Religion performed an absorbing function at every level; it sucked up, diluted and suppressed all forms of social unrest and innovative thinking. By attracting the most incisive critics of the system from the underclass and awarding them prestigious positions within its organisation, the Church maintained a flexible buffer between ruler and subject.

Intellectual life in the feudal period was diverted to the monasteries, where monks and nuns with literary talents were occupied with eternal discussion of insoluble theological dilemmas, and with the manual reproduction of biblical texts which were then stored in aristocratic libraries where they gathered dust; all with the ultimate goal of negating the critical edge of their intellect and directing it instead towards the maintenance of the existing power structure. If any monk showed himself to be interested in power and glory for his own sake, there was always the colourful cardinal's garb with which to placate him. Potential leaders of unrest were dressed in cowls or cassocks from an early age, thereby reinforcing the relentless suppression of the underclass.

All that was right emanated from an almighty God, and below him a holy hierarchy was constructed. God's appointed representative on earth was the monarch, whose religious and worldly authority alike were formalised in the laws that he himself caused to be produced and promulgated. The monarch, in turn, gave guarantees to the aristocracy regarding its privileges, by granting this dominant class a monopoly on the use of force. This monopoly encompassed both a right and a duty to God and the monarch to exert force to crush any attempt at armed revolt from the underclass. In exchange for this monopoly on force, which was also a concrete guarantee of control over the means of production, the sons of the aristocracy swore an oath of allegiance to the crown and assumed the role of officers when the monarch ended up in conflict with other monarchs and chose to go to war.

Feudal society was upheld by this alliance between monarch and aristocracy, anchored by the Church and embodied by the army. This led to the establishment of what every dominant class in every age has sought: social stability. The status quo was the principal and common interest of monarch and aristocracy. Every threat to the existing structure had to be smothered at birth. So an almost hermetically sealed system of society was created, where there were no opportunities to establish alternative centres of power from which the ruling structure could be attacked or even questioned. Consequently, there were no threats within the system itself; only a revolutionary alteration of the basic technological premises of feudalism could achieve noticeable movement and upheaval within the hierarchy, which in turn would eventually, and inevitably, lead to an entirely new society. In order for this to take place, there had to be a genuine paradigm shift.

The high status of religion during the feudal period did not rest upon any particularly widespread interest in existential matters among the population as a whole, but was the result of the aristocracy's intensive production of ideology, the purpose of which was to sanction the dominant class's unlimited right to own and preside over land. The religious message was the same everywhere: that every single piece of land had been given in perpetuity by God to one specially chosen family, whose inalienable right (and duty) it was to pass that land from generation to generation. The conservative religion of the Church, the laws put in place by the monarch, and the aristocracy's monopoly on force acted in concert to deny the peasants any effective means of questioning or opposing the feudal hierarchy and the forces in power. Land and inheritance were all-important to the aristocracy. The fact that property, rather than capital or knowledge, was passed down in inheritance within the family was the single most important prop of the dominant class. Property and the family name were consequently intimately connected; they were inherited as a package, and combined to form the most important symbol of feudalism: the family coat of arms, which was therefore imbued with exceptionally high status-value.

The question of whether or not there is a god is very much a modern one. The dominant classes have always been aware of the fact that God's invisibility is a problem. God's absence from the Earth creates a vacuum which must be filled by a representative to combat unease in society. There has always been a need for an ultimate arbiter whose advice could be sought in moral and existential questions. So whether or not God exists does not matter, as long as there is someone who can take his place. The most important thing for the aristocracy was clearly that this representative came from the right circles and served its own interests. Hence the appearance of the monarch in history. The monarch has always had divine qualities attributed to him, ever since the expressed divine lineage of the Egyptian pharaohs – which led to their having to marry their sisters. The status of the monarch should not be, and could not be questioned, because, like the right to inherit property, it was one of the corner-stones of religion. In this way, the monarchy and aristocracy co-existed in a balanced climate of religious and legal terror. No movement was possible; neither party could seriously question the authority or rights of the other without simultaneously calling into doubt its own privileges.

This balance of terror acted to suppress open confrontation, but created at the same time a perpetual cold war. As long as external threats seemed to be under control, there was a continual, low-level conflict between monarch and aristocracy. The monarch did his best to divide the many-headed aristocracy in order to be able to control it, whilst being fully aware that an aristocracy that was too weak would threaten his own position, because the peasants would be able to rebel under such circumstances. The aristocracy, on the other hand, sought internal unity in order to be able to control, as best it could, the isolated figure of the monarch, who was relatively weak in resources. Despite the fact that the combined power of the aristocracy was immense, its relation to the monarchy was problematical. It had to accept that it could not depose and appoint new monarchs at will, because that would undermine respect for the divine right to inheritance of land, and thereby weaken the position of the aristocracy itself. Therefore the aristocracy was forced to accept that inheritance of the throne was also sanctioned by God, which in turn strengthened the position of the monarch in the conflict between them. This explains the gradual increase in strength of the European monarchs from the late middle ages, relative to that of their aristocratic subjects, in line with the fact that their thrones, like the Egyptian pharaohs' and Roman Emperors' before them, came to be inherited.

The fundamental requirement for the belief system that emerged victorious from the Darwinian meme-war for religious power during the feudal period was that it must rest upon an all-encompassing, strictly authoritarian hierarchy of power. This meant that the monarch was able to stand above the aristocracy, and the aristocracy, in turn, above the peasants. God had to stand above the Church, and the Church above its congregation. The monarch's unlimited power over the judicial system gave him a weapon with which he could keep the aristocracy in its place. If ever he was in a tight corner, all he had to do was repeal the aristocracy's monopoly on force by making it legal for peasants to bear arms. At a stroke, this would have enabled the peasantry to depose the aristocracy and, if their rebellion succeeded, be rewarded by the monarch with aristocratic privileges, including the right to own land. This was the real-politik of feudalism. We can conclude that history, once all the palace revolutions and peasant revolts were over, always returned to a hierarchy where the monarch stood above the aristocracy, which in turn stood above the peasantry – so long as the central value in society was tied to agriculture. The power exerted by the peasants was restricted, in reality, to the few small corners of life which were so insignificant and peripheral that they were of no interest to the aristocracy. This can be compared to the way in which the impoverished masses of the capitalist era have been forced to pick at the scraps left over by the dominant class.

Just as the aristocracy once needed a monarch at the apex of the hierarchy of power, in his capacity as God's representative on earth, so the bourgeoisie, the dominant class of capitalist society, needed a representative for Man, the god of the new age. And this zenith of the capitalist hierarchy, this bearer of ultimate responsibility for making

Man both obedient and worthy of the role of God's successor, was the State. In the same way as the idea of God had died and been replaced by the idea of Man, so had the idea of monarchy died or been downgraded to a purely decorative function, and its place taken by the State.

Like the aristocracy before it, the bourgeoisie was highly conscious of the vacuum presented by the physical absence of the gods. This new god, Man, had no tangible shape either, but was more an abstraction, a representation of an idea, a phantom – which is why it was of vital importance to find a more or less credible representative which was reliable and could watch over the interests of the new dominant class. This new representative was the State in general, and parliament in particular; this was the voice of the people. A massive historical castling manoeuvre led to the displacement of the monarch: a law-making individual representing a fictional collective, by parliament: a law-making collective representing a fictional individual. Feudalism was replaced by and subordinated to capitalism, and the paradigm shift was a fact.

Christianity was the religion that was best suited as a guiding instrument for the aristocracy, and which therefore succeeded in the Darwinian meme-war between various different belief systems that was conducted during the introductory phase of feudalism. In the same way, a winner gradually crystallised out of the memetic Darwinian war between a mass of possible ideological mutations during the transition from feudalism to capitalism. The new paradigm demanded a new myth, and at its service it had humanism, ready to replace one fictional figure, God, with another, Man. Humanism was perfectly suited to the new circumstances, and became the perfect guiding instrument for the bourgeoisie, which now had to try to hold down the new underclass of capitalist society: the workers.

Like Christianity before it, humanism was a faith that presented itself as "the truth of the new age". Because God was no longer current, or, at least, not as unquestioned as before, Man was now at the apex of the value hierarchy, as holy as God had once been. Language, the capacity of the human species to think and express itself verbally, now magically accessible in mass-produced publications, was the starting point for a new, fictional structure where Man, as a result of this ability, was raised above and was of higher value than Beasts. But man was not born Man – because that would mean that humanism was an extremely poor instrument of power – but had to be educated and shaped over a long period, involving a great deal of effort, in order to reach the goal. For safety's sake, this was made a lifelong project. The State was appointed as a strict overseer, and the Market as the immutable yardstick.

This is the explanation for the creation of such phenomena as hospitals, prisons and educational establishments, as well as various political and academic institutions; their purpose was to define Man, and to correct undesirable deviations from the ideal "natural" state within the population. Capitalism has shown a sensational capacity for innovation in its constant production of new sicknesses, crimes and other defects among the citizens of its society, all of which require care and attention. The system has been so practically instituted that each new innovation creates a new market for therapists in white coats, and other experts, thereby granting them increased power. The perfect citizen has been one who has motivated him or herself to strive to imitate the ever more diffuse and unattainable ideal of Man; one who has been obsessed by the notion of living correctly, in accordance with the advice of the experts. All this to create a maximally effective producer during working hours and an insatiable consumer during leisure time; a citizen who gratefully spends every waking hour of the day on the constantly spinning hamster's wheel of capitalism.

Just as independently-minded individuals from the underclass were rendered harmless during the feudal era by being occupied with eternal theological questions in monasteries, the gifted children of the working class were placed behind school-desks, where their future was staked out by and within the social sciences. The entire project reached its culmination during the late capitalist period of the twentieth century, with the newly hatched idea of necessary "self-realisation", which led humanism into its final extreme phase, where every individual citizen was encouraged to become his or her own all-seeing moral policeman. With this, the supreme ideology of capitalism reached its climax. This explains why the bourgeoisie have defended holy humanism with such frenetic fervour against all attacks, real and imagined, and why it has been raised up as an eternal axiom, a religion. Almost every political party, from the Republicans in the USA to the East German Communist Party, has identified itself as humanist. It's the same old story: getting the ideology that legitimises power to appear "natural". It's simply a question of the bourgeoisie's own position as the dominant class of capitalist society, and how this is connected to humanism's position as the supreme ideology. In this light, capitalist society does not look like democracy in any real sense, but a humanist dictatorship.

The bourgeoisie sought what every dominant elite seeks: social stability, the undisturbed exercising of power, a social climate which is downright hostile to any imaginable alternative centre of power. In the same way as the aristocracy long before had sought to maintain the status quo, the new dominant class tried to create a closed social system; and, just as feudalism managed to absorb its own inherent contradictions, the immense tensions within capitalist society never posed any real threat to the hierarchy of power. It was only when the fundamental technological preconditions of capitalism underwent dramatic change that anything seriously affected it. It was time for another paradigm shift.

Because the capitalist system could not function without at least a semblance of a connection to the working masses, the entire legitimacy of the state rested upon the will of the people as it was assumed to be represented and expressed in parliament. The franchise was extended to avoid revolution. The idea that parliament represented the true will of the people was elevated to an axiom which could therefore never be questioned. A balance of terror was established between Ruler and Slave, and in order to consolidate its power the bourgeoisie made sure to create a mutual dependency between its own class and parliament.

Once the State had become an actor amongst all the others in the capitalist market, its existence dependent upon the tax exacted from labour and capital, parliament became subordinate to the capitalist system, and was incapable of questioning the fundamental basis of this without risking its own activities. Co-operation was paid for with money and privilege. Only advocates of a strong State were elected to parliament, because anything else was impossible by definition. The elected representatives may have defined themselves as right- or left-wing, but this was relatively unimportant; what mattered was that parliamentary debates never questioned the fundamental political idea of the bourgeoisie: Etatism. Governments may have changed colour from one season to the next, but the governing elite maintained its secure grip on power.

The political ideologies that have characterised bourgeois parliamentarism are really only different nuances of the same supreme ideology: Etatism, expressing a fundamental belief in Project Man, and in the historical task of the State to carry out this project. In order for them to exert power in peace and quiet, it has been in the Statists' interests to give the impression that every conceivable political force is contained beneath the parliamentary umbrella. This was what the dominant class had done during the feudal era: attract potential trouble-makers with offers of attractive positions close to the trough. In order to simulate a plethora of ideas and genuine contradictions, every little rhetorical difference between the parties' programmes was blown up to grotesque proportions. It was all a matter of making as much noise as possible in order to maintain the Statists' actual monopoly on the public arena.

Like every other collective, parliaments are based upon a common programme, at the same time as their activity is intended to conceal this basic fact. The strong State stamps everything that isn't Etatism or praise of a strong State as extremism, because this is part of the game. Great contradictions within Etatism are suggested in order to conceal the fact that its many different groupings are all co-operating to suppress the appearance of alternative oppositional forces. In actual fact all of them – conservatives, liberals and socialists, in both their democratic and totalitarian forms – subscribe to the same basic idea: that a strong State is necessary for the survival of a good, "natural" society. For many years the political parties were successful in their common manoeuvre for control of public opinion. But with the breakthrough of informationalism in the 1970s the situation was dramatically altered.

The most characteristic sign of a society on the brink of transition from capitalism to informationalism is a general medialisation. Before the breakthrough of interactive media in the 1990s, the media were characterised by a late-capitalist industrial structure. The leading media of this era, radio and, above all, television, were perfect instruments for the institutions of the bourgeoisie to transmit its message to the masses unchallenged – in the USA in the form of an oligopoly in which the largest industrial companies owned television stations, and in Europe principally in the form of state-run television monopolies. But with the pluralisation of the media – mainly as a result of the advertising industry's demands for a greater number of more specialised advertising markets – it was gradually released from the demand to play along with the ideological propaganda of Etatism. The media began to live their own life, forming the basis for a new power structure, and began instead to assume several of the characteristics of the informationalist paradigm's new dominant class: the netocracy.

As the medialisation of society has accelerated, representatives of the rapidly growing entertainment and media have become more willing to attack the interest groups that they perceive as blocking the path towards their own independence and growing power. Since the media are increasingly managing to exist independently of elected politicians, it is hardly surprising that politicians have been the main targets for sharp-shooting journalists. The media's strategy in their battle with the State is constructed around a fiction: the myth of the electorate's contempt for politicians. The core of this myth is the idea that the general population of late-capitalist society regards elected politicians as a group of corrupt crooks who are feathering their own nests at the expense of voters/tax-payers, and who consciously fail to carry out the tasks they have been elected to perform. Every nation has its own cultural variation of this myth; the politicians of every country are perceived as breaking the most sacred values of their particular culture. So American politicians are continually unfaithful to their wives, whereas their European counterparts engage in credit-card fraud, vote-rigging and tax evasion. So the citizens are turning their backs on politics in disgust – if the media are to be believed.

The problem is that this is a self-fulfilling prophecy: By constantly talking about this supposed contempt for politicians, the media have created a media phenomenon which, through its very existence, fuels demand for shocking reports. The concrete substance of all these reports is restricted to the fact that voter turn-out in elections in most western democracies has gradually fallen since the 1960s. Contempt for politicians has been elevated to an axiom, an irrefutable truth. Every politician or media-player who questions its existence is regarded as a heretic, an opponent who must be subdued, because he or she is standing in the way of the overriding ambitions of the media. It is not hard to see how public opinion and laws alike are constructed and shaped by the media. Politicians are producers, the voters consumers, and the media have appropriated the increasingly important role of curators of the political arena, and have therefore been able to exercise, according to netocratic principles, total control over the political process within informational society.

Everything to do with politics now takes place on the conditions of the media. The standard-bearers of representative democracy are in this respect completely powerless and can do nothing but adapt to the orders of their new masters. A political event which does not attract media attention is by definition a non-event. This means, naturally, that any last remnants of serious politics are confined to the shadows of media-driven dramaturgy, an apparatus whose main attraction is contempt for politicians. The fact that this is never questioned does not mean that it actually exists, merely that it is a "truth" which is popular within the circles whose purposes it serves: the netocratic media that have taken command of the public arena.

Let's suppose that contempt for politicians in its accepted form actually does exist. It ought to disappear or, at the very least, subside noticeably whenever a supposedly corrupt politician has been exposed and replaced. It would be a simple question of electing the right person to the right position. But this is not the case. There is no great difference in voter turn-out in elections whether the candidates are new and unsullied by scandal, or the same old faces. So this supposed contempt is not directed at any particular politician, hence it cannot be contempt for politicians that is the cause for the steady decline in voter turn-out. The explanation must be sought elsewhere. We can conclude that contempt for politicians is a myth, and that the people who created this myth probably have a vested interest in seeing it survive and appear to be "natural". This leads to a considerably more interesting question than the extent to which voters hold their politicians in contempt, which is the question of how this myth arose, and whose interests are served by its spread?

It is a question of power. If we compare the level of voter turn-out in different elections for different forms of political body, a clear pattern emerges. There is a direct link between power and turn-out: the more power is up for grabs, or, to put it another way, the more power is connected to the position or body to which people are being elected, then the greater the level of voter interest. This means that the crisis of democracy has nothing to do with a loss of faith in active politicians per se, but is mainly to do with a growing concern about the increasing impotence of politicians. The silent protest of a growing number of couch-potatoes is not directed towards politicians' abuse of power, but towards their loss of power.

This fact is, unfortunately, not particularly sexy, and does not fit in with media dramaturgy; it does not provide any attention-grabbing headlines, and it does not provide ammunition for a bout of populist mud-slinging. And, above all, it does not serve the interests of the media. So, instead, we are supplied with a constant stream of propaganda telling us about the righteous contempt that the people feel for the corrupt political class, which only serves to weaken the position of politicians still further, which in turn leads to a new round of gauntlet-running in the media. This process continues in a vicious circle, the inevitable culmination of which is the death of representative democracy, the complete impotence of politicians, and a hyper-real media dictatorship. This process is strengthened by feedback. Through the use of opinion polls, whose questions are obviously phrased by the media to serve their own purposes, the population is told what it thinks, and what it is "natural" to think. Then the media go on to show how adaptable politicians are adapting to this norm, or are allowing themselves to be adapted, and so the process goes on and on, ad infinitum. The investigations of the mass-media are, on their most profound level, investigations into the mass-media themselves. Statistics which purport to represent public opinion are actually the tools used by the media to manufacture opinion.

The sheer absurdity of this whole performance becomes obvious when the media start to judge candidates to all political positions according to purely media-driven criteria. The candidates' qualifications and competence to occupy the position in question are quite subordinate; the principal concern is that the candidates "give good media". The main consideration is whether or not they appear to be useful from a dramaturgical point of view, or, in other words, whether or not they can be exploited by the media in their constant search for new lambs to the slaughter, new sensations and scandals to fuel the headlines. The fact that attention is paid to this – the extent to which politicians are "media-friendly" – during the initial selection process illustrates the fact that the media are not content merely to reflect and cover the political process, but are actively directing and writing the script for it.

Political journalists are not concerned with politics as such, but with medial dramaturgy. Political questions are often far too complicated to come across well in the media, which is why anything that looks at all complicated is side-lined to provide space for artificial oppositions, symbolic questions and the private lives of politicians. Politicians willingly submit to get on intimate terms with the media's consumers – what else are they to do? To refuse would be tantamount to writing themselves out of the script of the political soap-opera. The boundary between politics and gossip is increasingly being erased. The politicians of the new age are like cabaret artists whose speciality is what the American sociologist Richard Sennett has called "psychic striptease". In other words, they make political capital out of their private lives. Intimacy attracts headlines which attract attention. The consequence of this growing phenomenon is that the feelings of public figures on this and that end up in the media spotlight while serious issues that demand time and thought are sidelined. But this increased intimacy also brings an increased risk for a media backlash. Being able to master the difficult art of personal exposure has become one of the most important keys for political success.

The process we are discussing is therefore created, controlled, reflected and "examined" by one and the same group – the media – and this system tolerates no scrutiny, analysis or criticism from the outside. Under the circumstances it is easy to believe that the goal of the entire apparatus is to serve the interests of the media. It is worth remembering here that those in positions of power within the media are not appointed by the people (whose interests are continually invoked), as politicians are – at least on a purely formal level. They are selected from within their own circles, hand-picked from internal networks, and given the task of serving the closed lodges and guilds of the netocrats. This is at the heart of the true crisis of democracy: the netocracy's assumption of power by stealth.

But because we are in a transitional phase there is still life in the old myths. The bourgeoisie is still cultivating the notion that representative democracy is immortal, and seems to have got grist to that particular mill from the collapse of the communist dictatorships of eastern Europe. The American social theoretician Francis Fukuyama has investigated the possibility that the historical process has come to a halt at the station of liberal democracy, but, at the same time, he has been unable to resist sowing the seeds of Nietzschean doubt: don't the equality and stability that crop up in every after-dinner speech actually imply an untenable stagnation?

Within the myth of representative democracy there is the idea of the excellence of civil society, whose dark-side is the fact that political apathy is a taboo subject. This cannot be discussed; silence shrouds the fact that those who possess the electoral franchise that so many earlier advocates of democracy fought and died for are declining to make their way to polling stations in ever increasing numbers, for the simple reason that politicians are increasingly impotent and powerless at the side of the arena in which the battle between well-organised interest groups is taking place. Obviously the media cannot be blamed – and cannot assume blame – for this flagrant lack of interest in politics: how could the media be in the wrong when it is they that are judge and jury in the case? Nor can the clients, the media's consumers, the people who must be entertained, be blamed, if audience figures are to be maintained. No, the ones who end up taking the blame are the naïve and defenceless elected politicians. Everything is the fault of politicians, but politics itself is supposed to be in great shape. This is the consciously constructed paradox that the netocracy is using to help it carry out its informationalist coup against the mortally wounded State.

The old myths of liberalism about the immortality of representative democracy and the excellence of civil society are based on the false assumption that these institutions are, once and for all, the best possible structures, which can therefore never be shaken or questioned. Nothing could be more wrong. The bizarre and tragic fact is that these same myths are managing to maintain their hold on a society that is undergoing a tornado of change. In actual fact the advance of informationalism has radically altered the conditions for the maintenance of society and democracy. Since the French social theoretician Alexis de Tocqueville enthusiastically reported home to Europe during the 1840s about how American democracy rested securely upon a network of interest groups, civil society has stood as the ideal and the precondition for a functional democracy. But the keyword here is network. When we enter a new historical phase, where social networks no longer have a complementary function, but instead dominate political development, the preconditions are dramatically altered. Tocqueville's civil society, the network of interest groups, has realised its potential thanks to informationalism, and has been transformed into an insatiable young cuckoo, a parasite on society, a reckless multi-headed monster, the jailer and overlord of representative democracy.

In the USA this network of interest groups, the lobby groups, have at least had the good taste to finance their activities themselves. In Europe, on the other hand, they are funded, via taxation, by the institution that they are fighting against, the State. If you examine this system from an informationalist perspective, you see how the new underclass is, in effect, financing the new dominant class's political lobby-groups through the tax system. As a result, representative democracy is effectively under attack from several directions at once. The rapid development of technology has resulted in the networks of interest groups becoming far more powerful, and their ability to exert political pressure has reached the extent where they have practically taken over and are actually in control of the political process. Forget the idea of one man, one vote. What matters now is being initiated into the right networks in order to be able to influence important decisions in any way that is not purely symbolic. The principal has become more like: one member of a network, one vote.

The new muscle of the lobby groups and advisory NGOs becomes even more powerful because of the fact that the working methods of these groups are perfectly suited to the general medialisation of society, which has created an unholy alliance between the interest groups and the mighty media. As soon as anyone from the marginalised political class tries to direct a specific issue in a particular direction, the affected lobby groups and advisory NGOs conjure up a speedy and pertinent expression of opinion, or, in other words: an artificial and well-directed rabble which, if necessary, goes on the attack both physically and virtually against the political project in question. From Greenpeace to legitimised white coats, from lawyers' associations to various netocratic mailing lists, concern and outrage can be arranged and produced to order. Time and again, the political process is paralysed, until in the end it subordinates itself entirely to the control of interest groups. These are in a position to dictate proposals themselves, in return for the poor politicians avoiding abuse from the media.

The only lingering function of the politician will be purely ceremonial; acting out a Punch and Judy show in the media, stamping documents which he has neither written nor understood on any level other than that of catchy slogans. In capitalist society the downgraded monarch had to be content with cutting blue ribbons to open shopping centres. In the same way, the power wielded by politicians under the netocrats will be limited to the use of their names to confirm and formalise decisions that have in practice been taken by other people whom the politicians have no real chance of influencing. But does this transition to an informationalist political structure necessarily mean that the principal of democracy is dead?

The explosive expansion of the internet has led many people to hope for a renaissance of democracy. Thanks to the fact that the technological preconditions exist for the citizens – at home, in the workplace, in libraries, etc – to express instantly their opinion on all manner of political issues, the net, according to this idea, could function as a sort of virtual parliament on both a local and national level. The net would do away with the need for representative middlemen and, therefore, not only function as the saviour of democracy in the face of media tyranny, but would also embody the fulfilment of definitive democracy, a liberal utopia. It's a fine thought. The problem is that the net does not make allowances for and does not recognise physical geography and its limitations. The identities around which people gathered and made decisions under capitalism were based upon preconditions that were directly connected to the old paradigm and which are now completely irrelevant. This means that the fundamental condition for a net democracy – that the group of people discussing and eventually taking decisions also has an interest in participating in the political process within a limited geographic area, a country, for instance – no longer exists. Why should people on the net engage in national political issues when the idea of the nation itself has collapsed?

The nation state is a fundamental part of the capitalist paradigm, and therefore has no credibility in informational society, where communication is built upon tribal identities and subcultures that are constructed according to completely different principals. This explains why wars between nations have ceased during the transition to informationalism, and have been replaced by conflicts between interest groups such as companies and pressure groups. People are simply no longer prepared to sacrifice their lives for a nation, just as the people of capitalism were no longer prepared to sacrifice their lives for feudal ideas like God or the monarch. The dream of a clinical war with no casualties, which has been nurtured during the latter days of capitalism (the Gulf War, the conflict in Kosovo, etc), must therefore be regarded as a direct result of the political opinion that defence of the nation or democracy is no longer worth any form of sacrifice. The difference between Hollywood's action films and postmodern war is negligible. War is only acceptable if it is reduced to a video-game with a predetermined victory for the "good guys". The nation has been reduced to a stage-set.

In the virtual world, virtual identities are important, which means that a new system for participation in the political process will have to be constructed, one which takes account of this fact. Parliamentary elections could certainly be conducted over the net, whereby everyone with the franchise could type in their personal code instead of going to the ballot box, but the democratic basis of democracy – broad debate in which all interested parties within a certain geographic area air their opinions on a specific issue – has now vanished. On the net everyone seeks out people of like mind, and constructs a new virtual space with them, free of the conflicts in and about physical space. No-one seeks out a group with which they have nothing in common. Ironically, the possibility offered by the net to find like-minded people, and to avoid people with whom we want no contact, also makes the net useless as a means of defending democracy.

The political structure which is developing on the net is fundamentally different from capitalist democracy. The netocrats' ever-present ability to leave the environment in which they find themselves, to move on if the current situation does not suit them, creates the preconditions for the growth of an entirely new and extremely complex political system: a plurarchy. The definition of a plurarchy in its purest form is a system in which every individual player decides over him or herself, but lacks the ability and opportunity to decide over any of the other players. The fundamental notion of democracy, whereby the majority decides over the minority when differences of opinion occur, is therefore impossible to maintain. On the net everyone is master of him or herself, for better or worse. This means that all collective interests, not least the maintenance of law and order, will come under intense pressure. A pure plurarchy means that it is impossible to formulate the conditions for a judicial state. The difference between legality and criminality ceases to exist.

This creates a society which it is almost impossible to get an overview of, in which all significant political decisions are taken within exclusive, closed groups, with no access to anyone outside them. Already during the late-capitalist period, the judicial system and the national banks of Europe and North America have left the democratic arena and become expert-led institutions, subordinate to the lords of the new order: judiciocrats within the field of law, and national-economic prophets. Political decisions are no longer taken through elections, in parliament or even in referenda on the net, but within closed networks whose members, like the members of medieval guilds, are selected from within their own ranks. Netocratic principals are replacing statist ones. Measurements of strength are replacing ideologies. The new ruling class that we are witnessing the birth of is not interested in democracy other than as a nostalgic curiosity. The ideological apparatus of the netocracy is now concerned with making this entire process appear "natural".

4.

INFORMATION, PROPAGANDA AND ENTERTAINMENT

In the beggining was not the word. That came later. And for a long time it had quite different meanings to those we use today. The Latin word "information" has two lexical definitions: firstly "representation", "depiction"; secondly "explanation", "interpretation". So the term relates to the intellect and our conceptual apparatus. When Cicero uses the verb "informare", he uses it to describe a sophisticated mental activity: to give something a form, to bring matter to life with a sort of active vision, ennobling it in the process. Form and matter were perceived as being in dialectic opposition to each other, and could be united, according to this way of looking at things, in a synthesis which has the character of an act of creation. Matter + form = life, according to the Aristotelian formula. This was of great interest to the thinkers of the age, but it was not possible to discern any direct link to the economy or to society in general. This sort of abstract reasoning was not something that greatly concerned the man in the street. The English noun "information" appeared during the Middle Ages, but did not attract much attention for many centuries.

A shift in meaning gradually occurred, unnoticed, but without the term thereby acquiring any greater status or particularly useful function in the eyes of the general populace; rather the opposite, in fact. During the first half of the twentieth century, when capitalism was in full bloom, information was something you looked up in a reference book or stored in files and archives: facts, details about one thing or another, either more or less interesting. It might be a matter of numbers, names, addresses, dates and so on. Information was handled by lowly civil servants, or in the less glamorous departments of larger businesses. There was no mention of "information theory" or "information technology", and a career in the management of information was scarcely anything to boast about.

Since then, there has been a swift and thorough shift in meaning, and the status of the term has risen dramatically. Information, formerly regarded as a dull but necessary lubricant in the production of goods and services, is now generally thought to be the hottest product of the entire economy. But that's not all: information theory has established itself as an overwhelming, intellectual meta-structure, the fundamental ideas of which have penetrated deep into a whole list of other important sciences, and which, to a large extent, determines the world-view which is taking shape within the new social and cultural paradigm. Technological information is today regarded as the very essence of the body of society, in the same way that genetic information is the key to biology. The economy revolves around information – indeed, life itself is a gigantic, endlessly complex and refined process of verification where information is stored and transported within and between us fragile individuals.

The shift in definition of the word "information" began in the USA during the 1950s, in conjunction with the evolutionary leap that occurred with the development of the earliest computers. The mathematician Norbert Wiener predicted a second industrial revolution driven by "thinking" machines which had the capacity to learn from the past and their own previous experiences. The central idea of this was feedback: that the machine used its own results as new data and made the necessary adjustments itself. Wiener regarded this feedback, and the "intelligent" handling of information, as the fundamental core of life itself. These ideas, with the help of the much-vaunted achievements of these remarkable computers, had considerable impact, first within the scientific community, and later in the general populace. They were the basis for the introduction of a whole new field of research at the intersection of mathematics, linguistics, electronics and philosophy, known as AI: artificial intelligence.

The mathematic theory of information completes the transformation of the term, which now denotes a purely quantitative measurement of communicative exchange. Prior to this, the word information may not have been a guarantee of quality – the information in question may have been extremely trivial and/or irrelevant – but now the word no longer assumes any position as to whether or not it denotes nonsense and/or sheer fabrications. Information is anything that can be transformed into digital code and communicated from a transmitter to a receiver through a communications medium, that's all. From an information-theoretical perspective, there is basically no difference between a scientific formula, a nursery rhyme, or a stream of clearly false election promises from a politician under pressure.

There is nothing new in science ascribing special meanings to commonly used words; this happens, for instance, in physics, just as it does in psychology, but this usually has no significance at all for general language usage, the contexts in which the words are used are far too exclusively scientific. But this instance is different, because it is connected to the paradigm shift and the generally increasing interest in information as a marketable product. The fact that information theory has been so successful, and has formed the basis of a succession of spectacular and profitable applications, has meant that the scientific usage of the word information has leaked into journalism and all sorts of popular- and sub-cultures. Information theory and economics alike are primarily interested in information in large quantities. The larger the better.

This means that the technology itself, the capacity for storing and communicating information, has become the focus of attention, while the actual content of the information is paid relatively little attention.

This is in the nature of the beast:: anyone involved in information theory is primarily concerned with the process of communication taking place in front of him, and therefore lacks any incentive to reflect upon the quality of its content, which is in any case extremely difficult to measure and build theories upon. What happens when this perspective comes to dominate the surrounding culture is that technological advances are generally seen as having the potential to solve all the social and cultural conflicts of our age. The solution is to throw information at the problem. This is the reasoning of those with blind faith, the ecstatically enthusiastic cheerleaders of the new dominant class.

This fixation with technology, the medium itself, is in its way perfectly understandable. The medium is the message, as Marshall McLuhan stated. Changes within information management and the development of communications technology are the main causes of social and cultural advancement. Without this particular insight, the development of society becomes completely incomprehensible. But information is not the same as knowledge, and as information is becoming the great staple product of the new economy, and the world is drowning in an ocean of unsorted media impulses, relevant and exclusive knowledge is becoming increasingly valuable. The advancing netocracy has realised this, in contrast to capitalism's clueless cheerleaders.

Any new, dominant technology as significant as the networking computer creates a new constellation of winners and losers. The winners consist of what Harold Innis, the great pioneer of modern communication studies, and McLuhan's mentor, calls a monopoly of knowledge. Those who control the new technology and its applications quickly accumulate considerable power, which inevitably and immediately leads to consolidation of the newly formed group, and a strong impulse to protect the interests of that group. For obvious reasons, it is impossible to expect that there would be much desire to make this exclusive knowledge accessible to a wider circle, which would devalue both the power and the privileges. One way for the winners to manipulate opinion is to claim that there are no winners, and that the blessings of the new technology will be spread evenly and fairly across society. This is when the cheerleaders, the great bulk of whom ironically consist of misled losers, come in useful.

Norbert Wiener's idea that the management of information is the greatest secret of life received the most prestigious confirmation imaginable when the biologists James Watson and Francis Crick cracked the genetic "code" in 1953, and learned how to "read" the spiralling text of the DNA molecule. The new biology has completely adopted the model and terminology of information theory, and it is practically impossible to imagine it existing without the shift to the new IT paradigm. We now realise that protein synthesis is an unusually refined example of information transference And the DNA molecule itself is nothing less than a perfect miniature computer.

The collaboration of the two disciplines had a positive impact on both their own authority within the scientific community and their status in the general consciousness. The benefits were mutual. Biology, now marching towards a brighter future, repaid its debt to information theory with interest by investing it with a sort of unfathomable, almost holy aura, a reflection of the mystique that came from the most hidden secrets of life. As a result, the metaphysics of the new paradigm began to take on clearer contours: the old mechanical world revealed by Newton, created and more or less regularly serviced by a creative god, was now replaced by a world of digital information, created by a virtual programmer.

Thanks to this, information in its new theoretical meaning also assumed a substance with almost magical powers in the world of pop-cultural perception. This has, of course, both benefited and been enthusiastically reinforced by the winners of the new paradigm and the new economy. But the transformation of information has been going on for a long time. The earliest electronic media, such as the telegraph, meant that information began to assume the form of a commodity, appearing as a mass of small "packages" that could be "sent" across great distances without any significant time-delay. The ability to receive information over great distances without delay often had economic and/or military benefits. This made new technology seem extremely attractive. But we must still bear in mind McLuhan's thesis: the medium is the message.

Speed and the abolishment of distance hastened the materialisation of information. The telegraph revolutionised the transport of information over long distances, and, as a consequence, a great deal of contemporary attention was paid to transportation and speed, while other aspects – such as interpretation, context and comprehension – were neglected. A piece of quickly transported information came to be regarded as something valuable in itself, quite regardless of what it meant or could be used for.

But the fact that technology solves one problem does not mean that all related problems disappear or become less pressing. Working out how to transport information over great distances does not mean that people know how to interpret and understand the information in a relevant context. What happens is simply that some things are left in the shade when the sun rises over a new paradigm. For every problem solved by a technology, at least one new problem usually appears. We have become used to thinking of motion as progress, but motion always has its price. Electronic information appears to be an isolated phenomenon, as individual cries, lacking resonance, in an increasingly fragmented world.

The most interesting thing, justifiably so, was the revolutionary fact that people could communicate back and forth from one end of the country to the other, whilst it was seldom questioned whether or not people actually had anything to say to the recipient. The materialising, atomising tendency is strengthened further when the information is, over the passage of time, increasingly sent in the form of images. Words still require a modicum of grammatical context if they are to mean anything at all, but images are assumed to speak for themselves. A photograph says all that need be said about the frozen moment; in the world of the photograph, everything is open to view. The present emerges in the light of a flashbulb, the past retreats into the shadows, and context dissolves into a thin haze. The value of information is high, but the knowledge content is decidedly uncertain.

When a trend-setting science such as information theory prescribes a quantitative definition of the concept of information, and when an increasing proportion of the economy is based upon large quantities of information, there is fertile soil for an almost religious information cult. The netocracy is in charge of appointing the priesthood. Technology defines all problem-solving as a question of the production and distribution of the largest possible amounts of information. Information is thrown at problems. The mechanical manipulation of information is believed to guarantee objectivity and untainted judgement – just like the camera and the photograph before it. Subjectivity is synonymous with ambiguity, unnecessary complexity and arbitrariness. It marks a deviation from the straight line, and is therefore the heretical antithesis of technology. According to the gospel of the information cult, the guarantee of freedom in our age, creativity and eternal happiness is the unrelenting, ecstatic flow of information.

But information can hardly be said to be a rare commodity today. It is difficult to suggest seriously that a significant number of the pressing problems of our age – social, political or personal – have their roots in a lack of information. The free flow of ever-increasing amounts of information, as Neil Postman and others have pointed out, is the solution to a nineteenth century problem that has already been solved. What we lack today is not information, but overview and context. The unrelenting and ecstatic flow of information is unsorted and unstructured: it must be sifted, organised and interpreted against the background of a coherent worldview if it is to be a source of knowledge and not confusion.

Multiplicity and pluralism are the highest honours of the new paradigm, obvious lodestars for the information cult. But mass and pluralism are in themselves highly problematical. How are we to choose? How do we discern useful information from nonsense and deceitful propaganda? In a dictatorship the apparatus of power strangles the flow of information with the help of censorship, thus making itself unreachable. But by flooding every channel with a torrent of incoherent information, the élite in power within a democracy – the best-organised lobby groups, the most influential media conglomerates – can effectively achieve the same result. Against any given collection of "facts", another instantly appears; against the report of any given piece of alarming scientific research, another more reassuring one appears. And so on. And afterwards everything gets back to business as usual. It looks like a vital democracy, but it is merely a spectacle for the masses. This torrent of information is thus no unforeseen phenomenon, and certainly in no way a fortuitous lottery jackpot for the citizens and consumers, but is actually a conscious strategy for maintaining social control. Powerful interest groups send out murky clouds of distracting information in order to maintain the secrecy of certain essential knowledge.

The overload of information and the lack of context are intimately linked, two aspects of the same subject. Together with other contributing factors – rapid urbanisation, the collapse of the nuclear family, the decline of traditional authorities – this state of permanent insecurity creates a value-vacuum, which is readily filled by all sorts of more or less reliable experts, all fully armed with even more information. Marriage, raising children, work-skills: everything today is in a state of great doubt. There are constant new decrees from the experts. The only sure thing, the only thing the experts have to support them, is modern science, and what that has taught us is that all knowledge is provisional – that all truths about the world sooner or later come to be revised. We know today that Newton, to take one example, was wrong about most things, but his model functioned, and that was what mattered. There is always something new on the way. Instead of The Truth, we have to make do with The Latest.

It was not only for Karl Marx that science took over religion's task to provide a meaning for history and existence. The only real articles of faith for bourgeois democracy are economic growth and scientific rationality. This means that all our political and social institutions ultimately rest upon a programmatically insecure set of values. Everything is in motion. Rationality not only has to act rationally, but must increasingly fulfil the functions of irrationality as well. This is generally the case for the whole of the parliamentary über-ideology, from right to left and covering every alternative in between: everyone sets their faith in rationality. Even the romantically inclined environmental movement refers to scientific rationality when it attacks science and its applications. Ecologism is, as the British journalist and author Bryan Appleyard has pointed out, "a way of turning science against itself". We believe in science, but what is it exactly that we believe in, when we know that something new is always on the way?

Experts are the new priesthood, our guides in spiritual and moral questions, mediating and commenting upon the very latest information. Statistics are the language of the new oracle, information presented as science. But what does it really matter when it turns out that the relation between the number of twins born and the number of female detectives has shown a noticeable change? What does it really matter that an immigrant from one continent seems more intelligent than an immigrant from another? And what does it mean to be intelligent? The masses, poor in knowledge yet over-informed, at the bottom of all low-status networks, are completely at the mercy of The Latest in its vulgar and trivialised form. The frequent showers of contradictory information have one single coherent message: don't trust your experiences and perceptions, listen to The Latest instead. But The Latest is quickly succeeded by The Very Latest, and it is practically impossible to imagine any information, any combination of new facts, that could affect the status quo to any noticeable degree: partly for practical reasons, because facts are so fickle and their rate of replacement so high, partly purely theoretically, because there is an absence of any context that is valid for the whole of the social collective, from which the implications of these facts could be determined.

The public, consisting of engaged citizens with a common interest in and responsibility for the general good, the classic precondition of democracy, never materialised in reality. Instead, the ever more vocal mass of the population and the ever more ambitious middle-class, which had so terrified the privileged élite of the nineteenth century, is being transformed under the relentlessly increasing pressure of information into a divided and ruler-friendly multitude of antagonistic special-interest groups. This means that all these statistical investigations, all this quasi-scientific social research, all of this torrent of new information which is so readily available to us in our efforts to make the world a little more comprehensible, is actually, to borrow Karl Kraus's aphoristic summation of psychoanalysis, "precisely the mental disorder for which it believes it is the cure". Or, in the words of David Bowie: "It's like putting out fire with gasoline".

The noble thought, which has its basis in Enlightenment philosophy, was however that facts spoke for themselves and that intellect would inevitably triumph. All privileges would be abolished and justice would prevail on earth. Thomas Jefferson was one of those who expressly spoke about the "diffusion of information" as one of the corner-stones of his political beliefs. The free press became the very embodiment of the virtues of liberty. A greatly increased spread of information via all the newly-established newspapers would not only provide sustenance to diverse progressive ideas, but also create the public platform where man could exercise the capacity for rational reasoning that was his innate gift, and the social participation that was his natural right.

This was largely how it worked. To begin with, at least. Information in general, and newspapers in particular, were an effective and justly feared weapon in the hands of the bourgeoisie which was seizing control from the old, temporally displaced regime. But once power had been seized and consolidated, there was naturally no burning desire within the newly established élite to carry on with the experiment. Freedom quickly became problematical once again, and all those people with their innate gifts and natural rights no longer appeared so regularly in the columns of the newspapers. If the eighteenth century's newspapers and pamphlets had reflected a genuine, revolutionary opinion that was critical of power, then the press during the nineteenth century was considerably more of an instrument of power with which the public was consciously manipulated and a general opinion manufactured. This means that it is still correct to regard the press, and information, as a weapon. But also that it is important to keep an eye on whose finger is on the trigger, and which interests this person represents.

The capitalist élite, like every other, sought to maintain the status quo. The primary task of its propaganda was to protect social, economic and political privileges in an age when privileges were no longer the fashion, and when increasing welfare and improved education were leading to demands for more rapid equality. Those in power sought, quite consciously, with the help of the PR-experts who were beginning to sharpen their tools by the turn of the last century, to turn the increase in welfare and the improvements in education to their own advantage and persuade the ambitious middle-class of the advantages of entering into a tacit but nonetheless effective pact.

The trick was to present the alternative, i.e. political change, as something extremely unpleasant. The message was that the newspaper-reading middle-class risked losing its hard-won gilt-edged life if the existing order was disturbed: that chaos and mob-rule threatened if the winds of change were set free. The trick worked. It was possible, it turned out, to manufacture opinion. PR-methods were built on science – what else? The PR-men regarded themselves as scientists. One of their trend-setting predecessors was the Frenchman Gustave Le Bon, who with great trepidation warned of what the easily-led masses might do, and how little they cared about laws and every other social institution; another was Gabriel Tarde, who, more cautiously, preferred to talk about the general public instead of the mass, and who had greater hope in what might be achieved with the help of the developing mass-media.

Tarde took up the Enlightenment idea of the debating public, "the grandiose unification of the common mind". But he also recognised the power and possibilities that lay within reach of the artful opinion-manufacturer. In modern society, the carefully-wrought message not only reached those who read newspapers, but everyone who spoke to all the people who read newspapers, i.e. pretty much everyone. According to Tarde, newspapers had created the conditions for a co-ordinated, public conversation where it was possible to pre-programme the right opinions: "One pen suffices to set off a million tongues".

The electronic mass-media offered even better opportunities to set people's tongues in motion. The printed word still requires a certain level of education in order to be understood, whereas radio, for instance, requires nothing but the flick of a switch and adequate hearing. This gave propagandists of all descriptions an opportunity to get inside people's homes in a way that was historically unique. Film, and eventually television, offered the opportunity to communicate directly through images. For someone looking at images, either moving or frozen, their perspective and the act of seeing are determined by the originator of the image. This opens what the critic and philosopher Walter Benjamin calls "the optics of the unconscious": a shortcut for the camera to bypass the intellectual censorship which comes into effect when we have to adopt a position towards a message on the abstract level of words; a direct channel into the individual's private dream-factory where unexpressed desires and fears roam about. Anyone who can manage to stay awake will experience a stimulating massage of his or her irrational inner self.

The theatrical presentation of democracy demanded steady direction if it was not to go off the rails: this was the main creed and business idea of the developing PR industry. Historical factors – welfare and education – made it necessary for the ruling and – naturally – responsible élite to develop a well-stocked toolbox of power so as to be prepared for all eventualities. General opinion was in constant need of tending and pruning, no undesirable weeds could be permitted to grow too tall. The masses could easily start to think a mass of foolish ideas. The medicine for this was information, but it must be administered by experts within the mass-media and the world of social-psychology.

With the help of precise science it would be possible, in theory at least, to fine-tune the intellectual and emotional life of the masses. This was known as the art of the Engineering of Consent.

When the bourgeoisie seized power from the aristocracy in conjunction with the previous great paradigm shift, information was, as we have seen, a valuable weapon, and when the bourgeoisie later sought to protect its position and privileges, information was an effective instrument of control. But now, when the netocrats are moving from their positions in the current transitional phase, information is a form of perpetual existential static-interference, making the figures in the image difficult to determine. It is no longer possible to achieve anything creative with information; the only result of a continued and unchecked torrent is an increase of mental pollution in society. The most characteristic quality of information in such great quantities are the great quantities themselves, and it is under all these layers of muffling and insulating padding that everything important that is going on is actually going on.

The PR-experts realised this at an early stage. These qualified forces are not engaged, therefore, with anything so simple as hawking information or opinions; they are re-creating reality itself. Today we live in a world where practically every moment of every individual's attention is exposed to the tender attentions of the PR-experts. All of life is packaged, stylised. Instead of doling out opinions about whatever happens to be the day's news, PR-experts set in play the very events that are the news. They do not provide a slant on reporting, but make sure that reality itself is pre-slanted for reporting so that general opinion is tinted in the desirable colour and nuance. Of course, lazy journalists can get ready-written articles from the PR-company to which they can put their name, but it is the performance of the event itself which is important.

This activity is carried out to an extent unimagined by the general population. One tangible example, taken from the sociologist Stuart Ewen, is how American opinion was programmed fundamentally and in good time for an eventual military offensive against Iraq in the early 1990s. One congressional enquiry that gained a great deal of media attention involved a fifteen-year-old girl from Kuwait, a hospital volunteer who testified that she had witnessed with her own eyes how, after the invasion of Kuwait, Iraqi troops had entered the hospital in Kuwait City, dragged premature babies out of their incubators and left them to die on cold corridor floors. This grim act of barbarism outraged the USA. No-one paid any heed to the fact that the girl in question remained anonymous, for the stated reason that her own safety was at risk. Only in retrospect, when the war was long over, did it emerge that the girl was actually Nayirah al-Sabah, daughter of the Kuwaiti ambassador to the USA, and that she could not possibly have seen the things she testified to having seen. It also became apparent that her appearance before the enquiry had been arranged by one Gary Hymel, vice-MD of Hill and Knowlton, one of the largest PR companies in the world, which counted the Kuwaiti royal family among its wealthy clients. Nayirah's testimony was part of a conscious and successful strategy, and was one of a great number of manufactured media events, all designed to direct American fury at Baghdad.

Sick and weak babies, dragged from their incubators and left to die on the floor: you need strong effects to attract attention and mobilise outraged emotions. Or any emotions at all. When access to information exceeds demand, the relationship is, for obvious reasons, quite the reverse where attention is concerned. Attention, pricked ears and focused eyed, is what there is a real shortage of in what is known as the new economy. This has to do with welfare and education. Sociologists from several countries have shown that there is a clear link between increased welfare and a higher level of education on the one hand, and a general sense of a lack of time on the other. Naturally, we have no less time than previous generations, or than exists in other less blessed regions of the planet: once again, it is multiplicity that is at issue here; the range of activities and leisure pursuits on offer is so much greater than before. There is so much that we would like to have the time to do within the time available.

Our patience with slowness is on the wane. We associate everything old with slowness, and we avoid everything old like the plague. Children and teenagers instinctively avoid black and white films when they happen to pop up during their channel-hopping, because they, not without reason, associate black and white with slowness. Anything that requires waiting is regarded as "dead time". The unspoken aim for anyone in tune with the age is to squeeze as much experience as possible into every single moment. Strong effects give the impression of a lot of experience, and entertainment is the cream that makes the sludge edible, and that separates it from other sludge-experiences. The one thing that we believe ourselves to know for certain is that life is short, and that our task in life is to fill it with as many different experiences as possible, as many kicks as possible. The winner is no longer the person who is closest to the king's arse (as under feudalism), or who has most money when he dies (as under capitalism), but the person who has experienced the largest number and the most extreme kicks.

Entertainment today is actually what information purports to be: a greater attention-magnet than any other, and therefore the economy's most significant driving-force. Development is approaching the point where every branch of the economy is coming to resemble entertainment, in great and eager strides. The shopping experience itself is being enriched with the addition of entertainment, a rich array of entertainment products is sold at petrol stations, and loaned out at libraries. Entertainment raises the consumer's sense of complete and appropriate consumption, and is therefore a decisive ingredient in the establishment of brand-names; it is, above all, "the e-factor", the entertainment factor, that makes one product appear more attractive than its competitors, and fuels turnover of goods on the shelves and on web-shopping pages.

Things have to be fun at all times. If people are bored, they go somewhere else at once, and consume something else. Las Vegas is not only the part of the world demonstrating the greatest growth, but is also trend-setting in terms of its intellectual climate. Academic stars invite their audiences to grand shows. And the need to be entertaining in the media has become one of the most important tasks for both politicians and businessmen. Everything suggests that we are now at the introductory stage of a phase in which information, paradoxically, is beginning to lose its prestige in the general consciousness. It is so easily accessible that it has become a logistical problem and an environmental hazard. If you conduct a search on the net and get four million hits, what are you supposed to do with all the information? There will come a day when the enthusiastic cheerleaders realise that they have been fooled. What is desirable is what is difficult to attain: an overview, context, knowledge. That is where power lies.

5.
Curators, nexialists and eternalists – The netocrats and their worldview

As a result of the ongoing paradigm shift and the transition from capitalism to informationalism, power is leaving the salons of the bourgeoisie and moving into the virtual world, where a new elite, the netocracy, is ready to take over. So who are these netocrats, and in what ways do they differ from their predecessors, the bourgeoisie? Where can we find them, and what are their distinguishing characteristics? What are their ambitions and strategies, their interests and values? How do they regard themselves and their social identity? And how is this new elite structured, what are the netocrats' internal distinctions and hierarchy? In order to approach this complex of questions seriously, we first have to understand the thinking and the circumstances that form the foundations for the progress of this new dominant class. We have to place it and its values in a historical context.

Ever since the earliest philosophy, western thought has been split into two main paths. We have chosen here to call these the totalistic and the mobilistic traditions, whilst remaining fully conscious of the objections that could be made against this division. The philosophy of the mobilist Heraclitus, for instance, inspired Plato, the totalistic disciple of the equally totalistic Socrates, and his fundamental concept of the world of ideas. But nonetheless, for pedagogical reasons, we have chosen here to focus on the differences between these two paths, instead of studying the overlap between them.

The totalistic tradition is characterised by the construction of the great system: a desire to find a single theory to encompass and explain the whole of existence and history. Within Chinese thought we find an equivalent to this ambition in Confucianism. Socrates, Plato and Aristotle are the three central figures within this tradition who have dominated western thought: their ideas have been nurtured and developed by great system-builders from Descartes and Kant to Hegel and the utopian Marx. But even Christianity and its theology, along with all of the political ideologies of the capitalist era, form part of the totalistic grouping. Both the church and the state as they have developed in our civilisation are to be regarded as totalistic institutions.

Totalistic thought is based upon the indivisible subject. The rules of philosophy are axiomatic, and are assumed as given from the outset. The ego is the basic building-block of this system, which means that the whole of existence is in orbit around this ego, like the moon orbits the earth. Thinking is situated within itself, and seeks to illuminate and regard existence from this assumed fundamental point. So observation is aimed from the ego, outwards at the world. Totalistic philosophy is interested in the relationship between the soul and the body (= ego and world). This is therefore basically dualistic. This way of thinking both presupposes and reflects upon its own productivity, and is fond of pondering moral and political questions. The ambition is to create a system which explains and provides a practical guide to life and the world. The fundamental questions revolve around man's identity: who is he, and what is his place in the world?

The totalistic question is a question in search of an answer. The question is the path, the answer is the truth, the truth is the goal, and a world in which all questions are answered is a perfect world, a translucent totality (hence the name totalism), a utopia made manifest. Plato claimed that this utopia already exists, that it is actually more real than the reality we believe that we perceive, which is merely a pale imitation of his hyper-real world of ideas. These ideas are the originals, out of our reach; the things that we perceive are of necessity only fallible copies. Christianity attached itself to this idea of the utopia that already exists, but in this case the connection to actual reality is more problematic. Utopia exists, but not here, and not now. The Christian utopia is in part a lost paradise, but also a coming state of heavenly joy; the world before the Fall, and the world after the Day of Judgement. In this way, when Christianity looks to the future, it is also looking to the past; its ambition is to reinstate what once existed.

For the totalists of later ages, in particular political ideologues, utopia is not a given fact but more a possible and desirable project. It is man himself who is gradually making this utopia a reality, first in thought, in the form of a vision, then in reality, through concentrated political activity. Since God is out of the picture, man himself must transform himself into God and become master of his own fate if he is to realise his utopia. The task of totalistic philosophy is then to stake out the path showing how this can happen. What unites all totalists is the idea of some sort of utopia which either has been, can be, or, at the very least, ought to be realised. This idea is connected to the idea of objective truth, an absolute, against which the state of things can be evaluated. The question is not whether life has a predetermined purpose, but what this purpose is.

One consequence of this is that totalistic thought is concerned with moral distinctions such as good and evil, black and white, high and low, right and wrong, useful and useless, etc. The goal is to place human thoughts and actions on different scales where these values constitute the poles. The task of philosophy is to determine these values once and for all, directing them from an imagined ideal state, from the absolute, or to create them with mankind's eternal requirements in mind. The intention is to lay a firm foundation for categorical judgements; every time we are asked what we really think about something or other, we are being asked to act as good totalists. Two and a half millennia of totalistic thought have created an almost incomprehensible spider's web of laws, rules, prejudices and collective obsessions.

One common aspect of all forms of totalistic philosophy is that thought itself has no value. The task of philosophy is ultimately purely instrumental: to render itself superfluous. When utopia has been achieved, totalistic thought will no longer be required, in the same way that a ladder has served its purpose for someone who has climbed out of a dark well. Until then, philosophy is a working tool, one productive discipline among many, providing man with something both useful and enjoyable. What differentiates the various totalistic threads from one another are the different absolutes from which their philosophy stems, which are, in turn, dependent upon what the desired utopia looks like.

This means that totalists are deeply divided on various central points: within this great arena there is plenty of room for religious and ideological wars and conflicts, of which history can provide many sorry examples. The enormous wealth of variety of content beneath the totalistic umbrella can easily deceive the observer into thinking that thought itself must be totalistically structured, and that there simply cannot be any other way of thinking. Both Kant and Hegel were aware of this and struggled with this question. This impression is reinforced by the fact that totalistic thought has dominated western culture to such an extent and for so long. Language itself has been occupied by the totalistic tradition, which has made the possibility of thinking in alternate ways more difficult.

Totalistic philosophy is almost obsessed by man's capacity to think abstractly, and, above all, it is fascinated by his ability to comprehend the fourth dimension, time, and to use this to view the world from both a backward- and a forward-looking perspective. For a totalist, it is natural to stress this awareness of time, so unique to man, and to remark constantly upon his consequently unique position in nature. For instance, man shares 98% of his genes with chimpanzees, but it is still this difference from his surroundings which is important. Life is a process of motion which has a decided direction, a clear beginning and a foreseeable ending. This means that the present is something secondary: it is the starting point in the past and, above all, the final destination in the future which are of most interest.

There is a central totalistic value in this. In this anthropocentric world-view everything is related to man and his needs: what is interesting about an object or an event or another creature is its similarities or usefulness to man, who is thus the measure of all things. The meaning of life is man himself, his wishes, and/or his individual salvation. The greater the similarity and/or usefulness to him, the greater the value. Taking Descartes, the first modern philosopher, as our lead, we can formulate the creed of humanism as follows: I think, therefore I am; and because it is I who think, it is also I who decide; and because it is I who decide, I shall force reality to bend to my will.

Totalistic thought is in all its forms strictly hierarchical. The definition of a man is a being who refuses to be an animal, and who is therefore of higher standing than an animal. When man is the measure of all things, the inevitable consequence is that man has an objectively true unique status. He is characterised by his ability to think and formulate abstract ideas, and this is what gives him a higher value. As long as man is alone in this, he possesses the highest value in creation, and everything around him is of a subordinate nature. This reasoning is clearly circular. Man is at the top of the hierarchy because he possesses the supreme qualities that make him man.

This is the crux of the great and insoluble dilemma of totalistic thought: how can a philosophy which refers to a pre-ordained hierarchy in which man stands higher than all other species avoid advocating an internal hierarchy within different species, and ultimately also between different human beings? How can it question an individual's claim to be allowed to rule over the rest of mankind? Perhaps he possesses these supreme qualities in especially large quantities? Plato claimed, famously, that the ideal would be to hand power to philosophers. Many other totalists have had the same idea. Some people are simply better suited than others to rule over the dull masses. And once the principle of hierarchy has been recognised and established, there is no end to the number of levels it can have. The less suitable may very well end up a long way from the top.

The fact that totalistic thought has dominated western culture to the extent that it has depends less upon the idea that it is intellectually superior, than that it has been, from the perspective of pure power, fantastically useful as a platform for social construction. During both feudalism and capitalism, every significant social force and myth took up position beneath a totalistic structure. Anyone with a utopian or eschatological vision has been able to call upon this structure, and in doing so has strengthened the legitimacy of the structure in return.

The dynamism in totalistic thought has made it useful both for supporting existing power structures and for criticising and toppling them. With reference to the totalistic ideal, God, the monarch, the state, democracy and other symbols of the elite in power have all been defended, just as revolutions and other projects for social upheaval have been legitimised. But now there is a spanner in the works. What is happening with the breakthrough of the informationalist paradigm is that this carefully constructed, universally recognised philosophical platform is under devastating attack from several directions at once. The bolts are straining, the joints are cracking. The grandiose totalistic model, the fundamental basis for the western social system, is imploding.

The transition from feudalism to capitalism was connected to paradigm shifts within both science and technology. It was astronomers and scientists like Copernicus, Kepler, Galileo and Newton who built the foundations for the new world-view. What the thinkers of early capitalism, the so-called Enlightenment philosophers, were concerned with was not so much genuinely new thought as patching and repairing the old, adapting traditional thinking to the new sciences and their revolutionary revelations about the nature of reality. The desire to ascertain an independent, objective truth and a centre of existence was still pressing, however. There was no ideological room to draw the philosophical consequences of this new world-view and its lack of a centre. This would have necessitated the abandonment of the totalistic platform, and it still had important tasks to fulfil.

The bourgeoisie had no problems coming to terms with the new perception of reality in which the earth was no longer the centre of the universe. In actual fact this indisputable fact, reinforced by the empirical evidence that could now be demonstrated, was an extremely useful weapon in the struggle against the old feudal power-structure. When it turned out that the earth revolved around the sun instead of vice versa, this meant that the whole authority of the old order was undermined and began to teeter. The whole of the old construction with God, the church, the monarch and the aristocracy had acted out its role, the players could be moved to the wings. But, on the other hand, it was impossible for the bourgeoisie to take this reasoning one step further and recognise that the new world-view also meant that man himself had been downgraded, and could no longer be the measure of all things and the central starting-point for philosophy. Such a conclusion was unacceptable, because it constituted a threat to the vital interests of the bourgeoisie. Man was God's appointed representative, and for this reason it was necessary, philosophically, to consolidate his unique position at the top of the hierarchy. For safety's sake, philosophy was fettered and marginalised by being designated a subordinate position as an eccentric deviant among the new humanist sciences. The capitalist era became a humanist dictatorship.

Three far-sighted and ground-breaking philosophers broke against the rules of adaptation, but their innovative thinking also had a high price. The Dutchman Baruch Spinoza was, quite simply, frozen out by his contemporaries, including his own Jewish community. His monism constituted a radical break with totalistic dualism. The Scot David Hume was forced to retreat and moderate his most radical ideas; and the first in a line of great German thinkers, G W Leibniz, found it necessary to camouflage the truly ground-breaking elements of his thinking – that the innermost essence of existence was motion rather than substance – amongst diverse advances within totalist thought that were more easily comprehensible and palatable to his contemporaries. In totalistic history books, Leibniz is more often regarded as a brilliant mathematician, "the last great Renaissance man", than a precursor of the Baroque within philosophy.

The breakthrough of the new world-view was incredibly powerful. For the capitalist power machinery, the empiricism of the natural sciences, i.e. the precision of mathematics applied to real experience instead of airy speculation about how reality ought to be arranged, appeared to be an extremely attractive attribute. The possibility of politics being associated with science and borrowing some of its credibility was regarded as potentially valuable. This would grant solidly-founded legitimacy to political power. In the long run politics could become a science in its own right. This project was accomplished in the 1800s when national economics, sociology and political science were established as academic disciplines. Hence an unholy alliance between politicians and academics was established, with the academics being left alone to manufacture truths that suited the ambitions of the bourgeoisie. But the advantages of this alliance were considerable even for science, which was guaranteed abundant resources and a wealth of attractive new tasks. The academic world gradually replaced old, outdated institutions like the court and the church as the recruiting ground for the political leadership-caste. Academic titles complemented a healthy bank-balance as the main attribute of the bourgeoisie.

Up to the end of the 1700s science was free, in so far as researchers could devote themselves, quite untroubled, to whatever they chose: to translate the Bible into the new national languages, to classify plants and languages, to study the heavens through telescopes. But thereafter political and commercial direction of the academic world was initiated, as a result of science being given great and prestigious tasks under capitalism: to provide protection for man's unique position in nature and install him in God's place at the top of the hierarchy. This is why the so-called humanist sciences were invented. The academic world was thus woven into, and became an indivisible part of the capitalist power-structure. The new, grand Project Man had been launched.

Since the capitalist system defined itself as thoroughly rational, there was no longer, ironically enough, any need for a philosophy that pointed out what was intellectually and morally correct, or what was irrationally and morally reprehensible. Questions like this were believed to be perfectly well handled by science, the market and representative democracy. Totalistic philosophy had, in other words, made itself redundant. Because totalistic thought was intimately connected to the abolished Aristotelian/Ptolemaic world-view, a fact which its practitioners refused to see, it could be reduced to a sort of therapeutic museum activity. To acknowledge the need of a new world-view would undermine the whole of their activity, and not many thinkers were willing to pay that price. Particularly not in a society where social exclusion meant rapid transportation to the proudest invention of the humanist sciences: the mental hospital.

At the same time, the bourgeoisie was not prepared to tolerate any philosophical alternative to totalistic tradition, since that was the basis of the humanist über-ideology that was sacred and above question. The tragic consequence of this deadlock was that philosophy under capitalism was controlled by a totalistic priesthood, a collective Gorbacheverie, that was doomed to wither away slowly and impotently, but which neither wanted to nor was permitted to abdicate. This, in turn, meant that the philosophical paradigm shift never happened: humanism meant, to all intents and purposes, a continuation of the old, the traditional; a secularised Christianity nailed up between St Paul and Aristotle. An all-encompassing revolution within philosophy would have to wait.

Only now, with the worldwide net taking shape and capitalist power-structures beginning to crumble, is the time right for totalism to be broadly questioned. The netocratic world-view is based upon thought which is certainly not new, and which itself can be traced back to Ancient Greece, but which has not hitherto been able to form a powerful alternative to the totalistic thought that has dominated philosophy up to now. We have called this alternative which characterises the thinking and perceptual world of the informationalist elite, the mobilistic tradition. It has its origins with the Greek philosopher Heraclitus, and has developed in near-obscurity throughout history, glowing faintly in the dark shadow cast by the dominant, totalistic tradition.

The mobilistic tradition is characterised first and foremost by a desire for universal openness. There is a desire in the subject to submit to the actual conditions of existence; to come to terms with existing circumstances, in order to use this position as the basis for attempting to improve the conditions imposed by fate. In other words, it concerns an attitude which is the complete opposite of totalistic philosophy: thought is here positioned out in existence, and looks at man from the outside. The ego is not taken for granted. Philosophy works from the world, towards the subject, an attitude which in eastern thought is found in Taoism and Mahayana Buddhism. The mobilistic question does not require an answer. It is instead a question which is constantly seeking the question which is concealed behind the question at hand. What the question expresses is a passionate desire for free and uncompromising thought, intellectual integrity; the answer can therefore never be anything but a cul-de-sac of thought, a comfort-blanket as a solace for philosophical cowards, a red herring detracting attention from the actions of underlying forces. The present is what exists, actuality is what is real.

Utopia, in its various forms, the dream of a controlling totality, is the main target of mobilists. Utopia is regarded solely as an instrument of power, demanding man's total submission and stopping him from thinking freely and living completely and fully in the present. Man is promised a reward in a more or less distant future, in return for giving up his freedom. He exchanges freedom for progress and the hope of participation in the coming utopia.

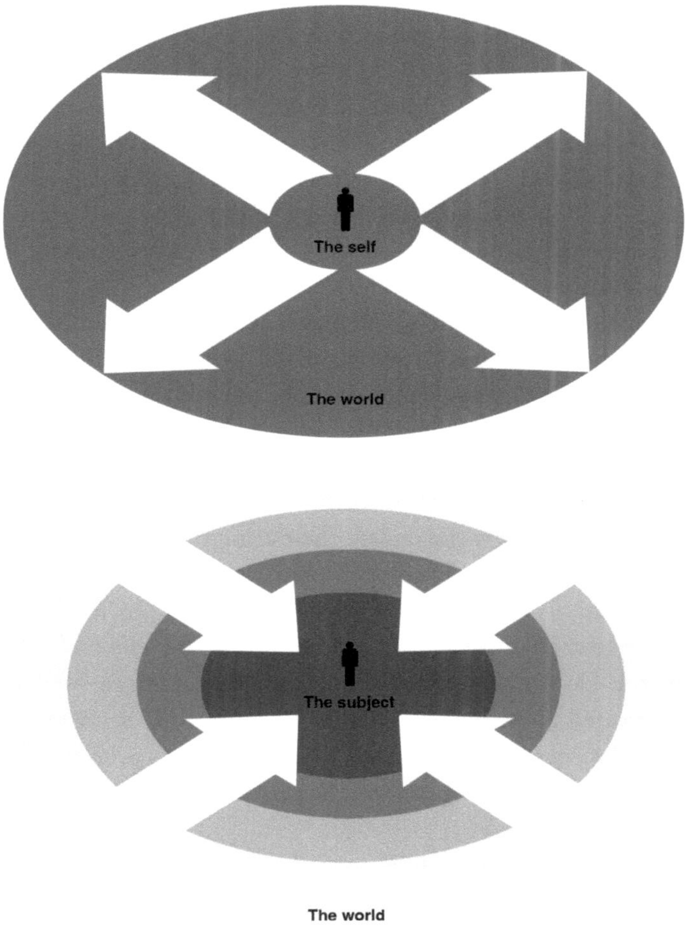

fig. 5.1 Totalism and mobilism

The path is lined with all the "objective truths" of totalistic philosophy, the axioms that the mobilist calls into question and identifies as the most cherished deceits of power: the ego, existence, dualism, hierarchy, laws, guilt, sacrifice, angst, memory, revenge, sympathy, progress, and so on. All these "truths" come together at the point where the reward is located, a reward for the self-assumed slavery that man is fooled, or allows himself to be fooled, or wants himself to be fooled, into suffering. One concrete example of this difference is when capitalists proudly renounce the present and postpone satisfaction of their needs to an uncertain future where this very postponement of life, this capitalisation, has a positive value.

Mobilistic philosophy rejects all of this and offers instead, as the only reward, the intoxication of freedom and the limited but real possibilities of the present. The primary task of mobilistic philosophy is that of a janitor: to clear the ensnaring intrigues of power away from thought. To uncover every attempt to objectify the hierarchies we are subjectively forced to construct in order to make existence comprehensible. This requires philosophers to formulate their criticism of power so that it stands above attitudes about "constructivity", because the demand for "constructivity" is power's demand that philosophy itself be made useful to power. A constructive critic of power is really an integrated part of the ideological power-system, because criticism of that sort is domesticated and harmless even at the moment it is formulated. It patches and mends. The task of the critic is reduced to protecting power by pointing out its failings, strengthening it against coming attacks, defending its position.

In the mobilistic tradition, thought has a value in itself. Mobilistic criticism therefore does not develop any dialogue with power, does not enter into horse-trading, but reveals the given "truths", "progress" and "rewards" as illusions and obsessions. It is, thus, "out of time". The demand for freedom also applies to the philosopher's relation to his own philosophy: thought must be entirely free. The very moment a philosopher proclaims ownership of his ideas, he is allying himself to the power that he is criticising. This is naturally problematic, because it means the mobilistic thinker can never be held responsible for the actual and practical consequences of his thoughts. There is a colossal risk in this, but also the enormous possibilities that are always part of mobilistic philosophy. You can never determine in advance where you are going to end up.

Ironically, mobilistic thinkers have always been able to count on the admiration of their totalistic colleagues, which has often taken the form of avarice. One example is how Machiavelli's revelatory reflections upon which strategies were effective in power-games at the highest level were appreciated and used as an instruction book by both politicians in Renaissance Europe and the leaders of late-capitalist businesses. Another example is how Nietzsche's fundamentally anti-fascist philosophy was turned inside out and used as an attack-weapon by the Nazis in 1930s' Germany. In this way, the mobilists' greatest admirers are often their worst enemies. Imitation is, famously, the sincerest form of flattery, but the imitator often misses or does not understand the very essence of what he is imitating. A totalist looks for benefits, not least to himself, and insists on logical strictness. Thought must be kept within the boundaries of language. This is why the acceptance in mobilistic philosophy of paradoxes, disinterested thought that is enough in itself, appears incomprehensible to him.

Even if mobilistic thought can be used by a cynic, mobilistic thinkers themselves often appear unfathomable and even ridiculous. This is the price of their refusal to join in the dance around totalistic truths. But with the arrival of informational society, the preconditions for thought are changing dramatically. This must not be understood as meaning that the informationalist paradigm is in any way "superior" or "more advanced" than its predecessors; reasoning in those terms means that we are still stuck in totalistic values which have been declared redundant by circumstances. On the contrary, informational society will, in many important respects, demand greater honesty of its participants. It will be more intellectually brutal than previous eras.

This honesty and brutalisation are central to an understanding of the netocracy and its values. Mobilism already offers these qualities, and is therefore rejecting – in meme-Darwinian fashion – totalism, which is collapsing under the weight of its discredited axioms. Mobilism can therefore, ironically, be said to be closer to the innate "truth" that it so firmly denies the existence of.

During the transition from capitalism to informationalism, a radical re-evaluation of man's self-image and world-view is being forced upon him. The altered circumstances demand new thought, but this new thought is actually not new, but something previously ignored, marginalised and misrepresented which is now in focus. Looking at developments from a biological/evolutionary perspective, socio-economic changes are favouring a mutation of thought which previously led an enfeebled existence. Voices from the periphery are beginning to be heard stronger and stronger. Ever since Christianity triumphed over Mithraism in the struggle over which belief system would replace ancient mythology as the state religion of the Roman Empire, mobilistic tradition has been confined to a secluded place on the edge of western thought. Free-thinkers like Lucretius, Machiavelli, Spinoza and Hume all recognised the limitations of the totalistic tradition, and attacked it to the extent that they thought advisable. But it was first in the 1800s, with Friedrich Nietzsche, that the mobilistic tradition seriously staked a claim within the philosophical arena. Immanuel Kant threw the door open, but it was Nietzsche who took the step into the new world-view.

Nietzsche rejected traditional totalistic questions about the meaning of everything, and the morals from the philosophical canon, and went instead straight to more demanding mobilistic questions about who it was who was saying what was being said, and why. With his "amor fati", love of fate, he ripped holes in the understanding that had governed philosophy since Descartes' time. He exposed the great totalistic project: the ambition for a totality of existence within philosophy, politics, science and art, a truth that was eternal and universally valid, to devastating criticism. Nietzsche rejected all talk about existence having an innermost core or objective purpose. There is, he claimed, merely an endless mass of conflicting forces that are constantly jousting with each other. It is basically pointless to speak of a fixed state of being; it is a question of a constant state of becoming. Existence is not something, it is becoming something, in the constantly shifting interplay of conflicting forces.

According to Nietzsche, all talk of morals was really about giving those in power an instrument with which to hold the masses in check, and above all, for the masses to hold the individual in check. He therefore called into question the entire totalistic Enlightenment project. Nietzsche claimed that it was not at all a question of creating a more open and better world for everyone, but rather the opposite: enclosing people within a sealed system where normality was the chosen lodestar and where bitterness and conformity were the predominant characteristics. The two main targets for his attack were Pauline Christianity and what he perceived as being its latter-day successor: humanism. Nietzsche regarded these forces as reactive and therefore reprehensible. He advocated instead his own ideal, the superman, whose actions are active and positive. He placed life and its immense wealth of variety above everything else. Free, uninhibited creativity was expressed by Nietzsche as what he termed the desire for power.

The single event that left the deepest impression on twentieth century philosophy was the student revolt in Paris in 1968. Students and organised communists met on the barricades in a unified revolt against bourgeois society. The large post-war generation that had taken over within French universities was driven, like flower power and the peace movement in the USA of the Vietnam War era, by the conviction that the capitalist system was bankrupt and was in need of a well-aimed shot to put it out of its misery. This diverse movement was led and inspired by a selection of charismatic figures, including the Marxist and existential philosopher Jean-Paul Sartre, who was strongly inspired by Mao.

But the student revolt failed. Fantasy never came to power. After a few months order was restored. This defeat led to a comprehensive re-evaluation of the accepted truths that had been cherished by the French intelligentsia. The working masses had not shown themselves at all interested in armed conflict, as the Maoist students and academics had imagined and hoped. The utopia had not been sufficiently attractive. During several hectic years the intellectual scene was radically transformed. At the beginning of the 1970s there was a general breakthrough for a new philosophy, with two Nietzscheans, Gilles Deleuze and Michel Foucault, in the vanguard. Mobilistic tradition thereby achieved a foothold in the academic world and began to exert an influence that kept on growing. Nietzsche conquered France and quickly expanded his empire.

Deleuze, Foucault and their many followers have been called, mainly by their philosophical opponents, post-modernists. This controversial title is based upon the idea that their criticism is mainly directed at the great project of latter-day totalism: modernism. As a counterweight to the dominant totalistic thinkers, Deleuze championed instead the pioneers of the mobilistic tradition – from Heraclitus, via Spinoza and Hume, to Nietzsche – but he also developed his own thought, which, according to many observers, Foucault among them, will go down in history as the most significant contribution to philosophy during the twentieth century. By unifying Spinoza's monism with Nietzsche's ultra-materialism, Deleuzianism makes a frontal attack on totalism's perception of the ego as a stable phenomenon, and on its dualism and dialectics.

Like Nietzsche, Deleuze sees existence as a constant conflict between forces moving in different directions; the balance of power between them is in a constant state of flux. The difference between the various forces is what interests him, and, from the point where the difference occurs, the point that Deleuze identifies as the singularity, it continues to expand unchecked at the same time as it constantly gives rise to new differences. This is thus a matter of a world-view in which existence cannot possibly be contained within human consciousness, since it is changing and expanding in all directions and at varying speeds, and in patterns whose complexity exceeds our capacity to comprehend. Consequently the totalistic ambition to gain a complete overview of existence appears completely absurd. Deleuze is completely uninterested in totalism's linear thought: of introductions, conclusions and totalities. His philosophy is instead concentrated on the centre of the mobilistic temporal axis, on the event, the feedback loop at the centre of things.

It is not the ego that produces thought but rather thought that produces the ego. When thought changes, so does the ego. There is no such thing as the fixed ego, the basic premise of totalism. So it is impossible to say that man in his capacity as sovereign subject can discover "the truth" by examining his surroundings. Instead we are forced to conclude that he largely constructs the truth that fits his purpose and circumstances. No truth survives outside the circumstances in which is created and where it fulfils a function. Totalism's search for "the universal truth" is therefore absurd. According to Deleuze, the task of philosophy is considerably more modest: it is to create functional concepts that help people to orientate themselves in existence, encouraging them to make their lives works of art. A new paradigm demands new concepts.

Deleuze, like Nietzsche, praises art. He sees philosophy as an art-form, connected to painting and music. He is concerned with the history of ideas, fascinated by how ideas gather in clusters in specific historical periods, only to disperse gradually afterwards. These ideas, like bodies, are in perpetual motion. He therefore called his thinking nomadic philosophy. Deleuze's ideal is what he calls "a body without organs", a complex structure which can be compared to an egg, where a mass of different factors are permitted to interact without the existence of any hierarchies between them, in order to create a whole which is greater than the sum of its parts. Deleuze is therefore usually counted as part of the mobilistic movement known as natural mobilistic philosophy.

The Deleuzian concept of a body without organs constitutes a passable parallel to Darwinism's genes and memes. In the meeting between Nietzsche, Darwin and Deleuze, the preconditions are in place for the first of three central figures in the netocratic system: the thinker whom we call the eternalist (after the Nietzschian concept of an eternal state of becoming). In the eternalist world-view, all existences, genes as well as memes, and the Deleuzian clusters, have a starting point, a singularity. From this singularity the phenomenon expands into eternity, giving rise, time after time, to new singularities, new complex patterns.

In the final trembling minutes of the capitalist paradigm the universe itself, through the consolidation of physics behind the Big Bang theory, has been transformed into a single, vast eternalistic phenomenon. The Big Bang theory is based upon the idea that the universe was created from a singularity from which it then expands for all eternity. In the eternalistic world-view this can be applied to all forces. And when a series of such forces work together as a Deleuzian historical cluster, a body without organs, then what eternalists would call a resonance-phenomenon or a feedback loop occurs. These temporarily blossoming clusters and resonances make up the nodes of civilisation. In the eternalistic world-view every single individual, the subject, is therefore a resonance-phenomenon rather than a fixed ego.

When singularities which have sprung from each other end up in a confined space, they are bound to meet sooner or later. The patterns which are then conjured up are an exact parallel to the system of contacts that arises in the development of a network. This is where the eternalistic world-view meets reality in netocratic society. The world is perceived as a single organic network, the all-encompassing net, where the clusters of genes and memes that arise are the nodes of the network. If eternalists are the interpreters of this reality, then the actors who appear at the nodes, the entrepreneurs, are another category of netocrats: the nexialists (after Latin "nexus"). The path to these nexialists, or the connection between them is managed by the third and most powerful of the netocratic categories: the curators. It is the curators who point the way for the nexialists, while their mutual world-view is constructed by the philosophers of netocratic society, the analytical eternalists. In the interaction between these three roles, netocratic society is created. If we make a general comparison with capitalism's power hierarchies, we could say that the curator replaces the politician, the nexialist replaces the entrepreneur, and the eternalist replaces the academic in netocratic society.

If Deleuze has become the supreme mobilistic philosopher, Foucault has become the great mobilistic historian, or rather its archaeologist of knowledge, as he himself preferred to be called. To Foucault, nothing in society is "natural": the word itself is an expression of the totalitarian ambitions of those in power, a desire to do away with everything undesirable by declaring it "unnatural". The central aim in social conflicts is to conquer the power of definition. Foucault works from the marginalised groups of capitalist society, the outcasts and their desires and needs, the excluded, as he defined himself. According to Foucault, the task of the philosopher is to silence power, to free man from the enslavement of utopias. The goal is for the weak man to be able to express himself.

Instead of a democracy, where the majority constantly overrides the minority, Foucault advocated a plurocracy, a society where everyone could make decisions for himself, but is not allowed to decide over anyone other than himself. What Foucault did not foresee was that this plurocracy would largely be realised by informational society's technologically driven transfer from democracy to plurarchy (plurocracy is an imagined political model, whereas plurarchy is a social state). Deleuze and Foucault were both fascinated by the electronic media, and showed an almost intuitive understanding of the changes and new possibilities that would follow in the footsteps of technological development. Their thinking has many aspects in common with our own analysis of the informationalist paradigm, and is highly applicable for anyone hoping to understand how both the new elite, the netocracy, and the new underclass, the consumtariat (CONSUMer proleTARIAT), see themselves and the world.

One example of a typical netocratic dilemma is the recurrent choice between exploitation and imploitation. Suppose two netocrats meet on a far-off island with picturesque ruins and beautiful beaches, but with no tourist industry at all. This is a typical netocratic destination, a perfect place for someone who practises tourism in the form of imploitative consumption. When the two netocrats are sitting on their sun-loungers, sipping cold drinks at sunset, they are faced with the question of whether they should keep the island a secret and only tell their closest friends of its existence, or build hotels and an airport and then market the island as a destination for all the tourists of the world: put simply, should they improve it and then sell it to the highest bidder.

If they choose to keep the island secret, they will be following an imploitative strategy; if they choose to make a profit from their discovery, they will be following the opposite, exploitative strategy. The difference between netocrats and classical capitalists is that the netocrats have these two options. Knowledge of the island has such a high value to the netocrats, and profit such a relatively low one, that exclusivity could well weigh heavier than economic profit. For the capitalist there is no choice. For him the accumulation of capital is the central project in life, a project compared to which everything else is subordinate. But the netocrat does not share this view. Conscious of the fact that his new-found paradise would lose its unique aura if it was exploited, the netocrat can choose, thanks to his independence from and lack of interest in capital, to imploit the island instead: to keep it secret and reserve it for the pleasure of himself and his netocratic colleagues.

Characteristic of exploitative consumption is that payment is made with capital. This is different to imploitative consumption, where money is largely uninteresting, and where it is a matter of knowledge and contacts instead, belonging to the chosen few who possess exclusive information. Entry into this circle cannot be bought with money, in the way that the nouveaux riches used to buy status with the profits of their businesses, but can only be achieved if you yourself have knowledge, contacts and exclusive information to offer in return. This means that for the old dominant class, the bourgeoisie, and the new underclass, the consumtariat, exploitative consumption is all that is on offer. Imploitative consumption is reserved for the netocracy.

The same dynamic forms the very basis for the power structures of informational society. A common misconception among the information theorists of late-capitalism is that the network's transparency will result in a more open society with full democratic visibility on all levels and where all participants have the same possibility to influence and the same access to information. But this reasoning should be regarded as palliative netocratic propaganda. This democratic utopia is an sign of rationalistic wishful-thinking and is based upon the misunderstanding that the internal dynamic of networks, on the micro-level, is automatically transferable to society at large. It is not that simple. What is valid within a network is only valid there, and says nothing about the dynamic that pertains on the macro-level, between the different networks, or, in other words, for virtual society as a whole.

Informational society is highly dominated by power hierarchies. These, however, are not constructed in the traditional way – with individuals, companies or organisations – but with membership of networks. At the bottom of this power-pyramid we find, once again, the consumtariat, trapped in a network of exploitative consumption where anyone can become a member. This base network is characterised by the fact that its main activity, directed consumption, is regulated from above. The system prompts desire with the help of adverts and then provides sufficient payment to maintain consumption on a level deemed suitable by the netocracy. This is hyper-capitalism sublimated to the level of sedative: the main concern is not to maximise profit but to prevent riots and virtual violence directed at the netocracy. Above this broad basal network, constantly-renewed and smaller networks are constructed, all competing with one another. These function according to capitalist principles (the traditional golf-club is a suitable model). Only those who can afford it can gain access here. But at the top of the hierarchy, only those who possess attentional value gain entry, in other words: those who have contacts and knowledge that are in themselves valuable to the network. It is here, at the top of the hierarchy, that we find the dominant netocratic class.

In this calculated way a merciless power structure of networks is constructed, where the most exclusive network, to which only the uppermost netocratic elite has access, is at the top. Family names mean nothing here, as they did under feudalism. Wealth means nothing here, as it did under capitalism. The decisive factor governing where in the hierarchy an individual ends up is instead his or her attentionality: their access to and capacity to absorb, sort, overview, generate the necessary attention for, and share valuable information. Power will be more difficult than ever to localise, not to mention how difficult it will be to watch over and influence. Social climbing will become even more complicated than it was under capitalism, the unwritten rules even more complex and inaccessible.

The interest of the netocratic powers in exclusivity and secrecy, combined with the increasingly rapid pace of change within society, means that the rules of netocratic society will be impossible to formalise. As a result of the fact that netiquette is a matter of what is unspoken rather than written down, of the intuitive rather than the rational, it will be the only possible set of rules for social relations in a society characterised by discretion and mobility. Laws and regulations of the traditional western variety have essentially played out their role. The ironic thing in these circumstances is that the netocracy is achieving its advantage over both capitalists and the consumtariat by making use of the virtues of mobilistic philosophy. In high-status networks there is no room for boasting and self-assertion. Instead, openness and generosity are what is most prized there.

It is, paradoxically, the netocrat's ability to think beyond his own ego, to build his identity on membership of a group instead of individualism, on electronic tribalism instead of mass-medial self-assertion, that lead to him understanding and being in control of the new world that is developing. Anxious tinkering with one's own ego, overplayed individualism, is instead characteristic of the new underclass. It is this very inability to see beyond their own ego and its desires that means that the underclass will remain an underclass. Much-vaunted self-realisation is becoming a form of therapy which is keeping the old bourgeoisie and the new consumtariat occupied with private problems instead of interesting them in questioning the new order. Anyone who "believes in himself" is by definition a hopeless loser in the society dominated by the netocracy. In important networks, no-one has the time or inclination to listen to a self-obsessed ego. Networking itself, the feedback loop and social intelligence are at the very heart of the netocracy.

6.
Globalisation, the death of mass media and the growth of the consumtariat

According to mobilistic philosophy every force has an opposite; every movement meets a resistant movement which offers a greater or lesser amount of opposition. In speaking of a new dominant class, the netocracy, we are presupposing the existence of its antithetical shadow, a new underclass which adopts the position and role occupied by the working class in the capitalist paradigm. The question is: which qualities are going to characterise and therefore define the new mass of people who will be the subjects of the netocracy? The defining characteristic of both peasants and industrial labourers was that they provided their masters with physical strength. Technological developments towards ever more refined and automated production processes have drastically reduced the significance of the human factor within the manufacturing industries; the physical labourer has either migrated to the service industries or become specialised in the supervision of sensitive and complicated apparatus; a labour mannequin, to borrow one of the philosopher Jean Baudrillard's phrases.

In other words, the underclass no longer consists of labourers in the accepted sense of the word. The defining characteristic of the new underclass is not its function either as raw material or as an expense for the enterprises of the dominant class, but rather as consumers of these enterprises. The main point here is not what the underclass produces, or even whether it produces anything at all, but, above all, what it consumes, and, even more importantly, the fact that it consumes at all. The proletariat of informationalism will first and foremost be a proletariat of consumption, or, as we have chosen to call it, a consumtariat. The defining characteristic of this class is not that it plays a subordinate role in production, but that it consumes on the orders of those above it.

In the capitalist paradigm paid labour was the basis of the entire economic system. This means that paid labour has been of vital ideological significance. To be productive was the very definition of being a successful human being. Talent was defined as the ability – and a quantifiable ability – to produce goods and services which could be sold in the marketplace. The combined economic value that the market placed on all waged labour – regardless of the extent to which this ended up in the workers' pockets as wages, in the investors' pockets as profits, or in the Treasury's pockets as taxes – has been the measure by which entire national production has been calculated. This is the only aspect of human activity which has seriously interested the bourgeoisie.

The overriding concern of every individual capitalist has been to maximise profits, which has often resulted in a hunt for unnecessary costs and the redundancies that are a natural consequence of this. But capitalism itself, both in practice and as an ideology central to society, has sought instead to maximise the number of paid workers, and involve as many people as possible in the apparatus of production. The state and the markets have therefore often been mistakenly regarded as opposites, particularly during the Cold War, but have actually comprised two separate but nevertheless mutually dependent pillars of the organic structure of capitalism, regardless of what the political system may have been called. The demands of the state for increased production and the demands of individual capitalists for increased profits have coalesced into one single aim, into a marriage between the state and the labour market which even its participants were unable or unwilling to hinder, a symbiosis that was impervious to other forces. This unholy alliance, this forced alignment of collective and individual wills has been both driven forward and defended by the overwhelming goal of capitalist ideology: to achieve maximum growth in the economy for the sake of growth itself. Different political ideologies have actually only disagreed on the best way of reaching this common goal.

Behind this overriding ambition is concealed the philosophical utopia of rationalism: all human needs, which are assumed to be constant, will be fulfilled by steady, continual growth. Once this point is reached the rationalist utopia will have been realised. With a common and abstract goal for the entire body of society, comprising all political ideologies and commercial forces, no-one need think for themselves any more. The supposed conflicts of the late capitalist era between individualism and communitarianism are best regarded as political theatre, because there have never been any fundamental differences between the various political programmes. Libertarian individualism has never, for instance, been a matter of freeing the individual from an enforced collective identity, but has been interested in the individual entrepreneur's demands for lower taxes ("tax is theft"), and higher productivity within the state apparatus ("the night-watchman state"). On the other hand, however, the fact that capitalism has forced individuals to set aside generalities within their own identity in favour of a specialism demanded by the system has never been called into question.

The dominant role of the state during the late capitalist era has manifested itself in two ways: one of them European, where the state is one of the leading players in the market, and the other American, where big business exerts strict control over politics by using the carrot and the stick approach. In both instances the result has been that the political and economic sectors have practically merged: politics has become economised, the market has become politicised. Political economy and economic politics have become one and the same thing: the rhetorical ritual of rationalist religion. Neither State nor Market has been able to accept any form of human activity outside their collective construction of civilisation. Measuring economic growth has been capitalism's means of quantifying the extent of civilisation itself. In the end the bourgeoisie imagined that it had reached its goal: a universal coalition behind the idea of the social body as a self-perpetuating, well-oiled, self-improving production machine. The problem is that capitalism has not really been victorious but has played out its historical role. The bourgeoisie has a new problem now: a new and developing dominant class with completely different ideas than those encouraged by the lords of the capitalist paradigm.

One consequence of this development is that what we call globalisation is actually two entirely separate phenomena. The capitalist globalisation process is a purely economic phenomenon and is directed towards continued specialisation and diversification. Increased competition is not visible so much in the form of direct confrontations as it is in the division of every market into several smaller, ever more specialised sub-departments. Every player, both individuals and entire cultures, is forced to distil those qualities which are in demand by the particular niche of the market holding sway at that moment, to the detriment of every other speciality and overview. This leads to the development of an increasingly tightly connected, finely-meshed network of mutual dependency. We are talking about a mercantile balancing act, an enforced act of co-operation in the shadow of the threat of the collapse of global trade. This aspect of globalisation is directly connected to the old paradigm, and it is necessary to distinguish this phenomenon from the parallel globalisation project which is part of the new paradigm.

The capitalist globalisation project implies a link-up between the most effective, and therefore most profitable, production apparatus with the wealthiest, and therefore most willing to pay, consumption apparatus. This arrangement is aimed exclusively at facilitating the traffic of goods, services and capital across old national boundaries. Interest in the freedom of movement of individuals is limited to their capacity as labour. It does not automatically follow that there is any general interest in individuals and/or their ideas. Ideas are only interesting in their capacity as products protected by copyright laws, or, in other words, as tradable commodities.

This project is a consequence of new technology: the extreme mobility it offers, its speed and diffuse locality, all qualities which in combination mean that the market is freeing itself entirely from traditional laws, rules and limitations. The basic idea of the project is to confront the politicians of the world with a fait accompli, and drive through a global market free from all tariffs, regulations and, as far as possible, taxes. The purpose of this is, of course, to maximise profits. Because the potential of new technology is so well-suited to capital, the already well-advanced globalisation is forcing the political establishment to retreat. This is expressed in various ways: either as a resigned and ultimately unsustainable isolationism, or, more usually, as an entirely new note in the rhetorical repertoire. Suddenly all the democratic socialists of the world are converting on the gallows and uttering uncompromising paeans of praise to free trade and classical liberalism. This sort of political volte-face should be regarded as a last desperate attempt by the professional political class of capitalism to cling to the last illusory remnants of power. At the same time it makes politics look important and relevant, so that it can "give good media".

The netocratic globalisation project is something altogether different, more of a social phenomenon, based upon the inherent possibilities of the new technologies for communication and contact across great distances and between different cultures. If the great goal of the capitalist is to maximise profit in order to eventually retire and nurture his individual identity, then the netocrat's great aim is to improve and facilitate communications between himself and all the strange experiences and lifestyles which new technology brings within reach. The netocrat seeks out the universal in the global arena, he wants to come up with a universal language, through the use of which he can experience all the exotic impulses he is longing for.

We do not mean to imply that one project is better than the other: it is simply a question of two different forces with different aims within two different systems. In both cases it is a matter of electronic colonialism;

economic in the case of the bourgeoisie, cultural for the netocracy. What is interesting from our point of view is the possibility that these forces might not run parallel in the future. When the capitalist project develops in a direction contrary to the structures of the network, which are controlled by the inherent characteristics of technology, it becomes more difficult to control, particularly for the capitalists themselves, who slowly but surely will lose their power to the netocrats. The netocratic globalisation project, on the other hand, cannot possibly fail, and will reward its participants, the netocrats themselves, with increased power.

The late capitalist age is suffering from schizophrenia. It survives, and has always survived, through adaptation, but is obsessed with control, totality and zero risk gambles. If capitalism surrenders the nation state, for instance, this would not be evidence of any new thinking in principle, but merely recognition of the fact that the highest instance of control must be transferred to a supra-national, federal level. Totality expands but the controlling and guiding ambition remains the same. The agreement between capitalism and the netocracy is not merely concerned with differences in background, lifestyle and attitude. The paradigm shift is about a fundamentally altered world-view. History is losing its predetermined direction, utopia is disappearing. The only way forward is no longer the only way; from every point of departure there is an infinite number of possibilities in the form of untrodden paths. Totality, rationalism and orchestrated collectivism are collapsing under the pressure of the virtual world's diversity. The netocracy is replacing the bourgeoisie, dragging the consumtariat along behind it.

The capitalist world is by definition economic, and the choices we are confronted with each day are primarily economic in character. It highlights only those activities which can be registered and measured in economic terms. Consequently capitalism has made money from every possible market and has turned every conceivable resource into a commodity. This enforced economic exploitation of everything it can find is called by the Australian philosopher and social-theorist Brian Massumi "the additivity of capitalism". The state and the market are united in their hostility towards activities which take place outside the economic sector – housework, various forms of unpaid voluntary work, etc. This hostility explains the comprehensive transformation of such activities into controllable and taxable paid work. Instead of parents helping one another with the supervision of children, this activity has become a profession practised by educated experts in return for monetary payment. Professionalism expands, and no activity is too simple to escape the attentions of experts. It is not the task itself which carries status, but the career.

When parents look after one another's children instead of their own, their work can be taxed and included in the state's statistics. In this way there is growth; the parents are included in the production apparatus, they can be registered and become the objects of state legislation governing the care of minors, and everyone is happy. This combined redefinition and redirection of various types of work is a typical example of how growth can be manufactured with a few simple manipulations in the late stages of capitalist society. This also shows that it is not only profit-maximising companies but also, to a similar extent, the welfare state which is showing acute signs of accelerating additivity. Here we are talking about an "economism" whose claims on hegemony have never seriously been challenged.

Capitalism has simply been overwhelmingly successful; it has functioned, which has led to its ideology appearing to be self-evident, elevated above all criticism and therefore almost invisible. Capitalism's appearances under various names have camouflaged its actual monopoly on power by continuously airing political disputes between party-lines which all co-exist nicely under the same meta-ideological umbrella. But there is a reason why capitalism has succeeded so well, and that is that it has been so well-suited to the existing technological and social preconditions. Now that these are undergoing drastic change, everything is suddenly in question. With the breakthrough of informationalism, the previously unassailable position occupied by capitalism is under attack from several directions at once.

The task of the working class was to work at low cost. It has been in the interests of both the bourgeoisie and the state to keep wages as low as possible, but also to avoid violent clashes and if possible to maintain peace in the labour market. Strikes are the central issue here. By giving the workers the right to strike – that is to say, to protest peacefully against low wages – the bourgeoisie was able to guard its monopoly on power. The working classes were effectively disarmed by this, and at the same time it became possible to ascertain exactly, and under relatively orderly conditions, the point at which low wages and bad working conditions threatened to boil over in discontent. Businesses could maximise profits if wages were fixed just above this critical level, and all parties were expected to be satisfied with this. Revolution was postponed once more, and at the lowest possible price. This ritual was repeated each year with a good deal of commotion.

This entire procedure, and the whole of classical capitalist mythology, was permeated by and largely based upon a vague but nonetheless grand promise: the bourgeoisie's undertaking to use material improvements to raise the working class to its own level and thus peacefully achieve the Marxist utopia of a classless society. There was no need for an immoral and violent revolution, the working class just had to grit its teeth and apply itself, and it would be richly rewarded. In this way, with the help of a semblance of common interests, the elite could build alliances with the spokespeople of the working class. The elevation of the working class was a great project of cultural revisionism, and fulfilled its task perfectly: pacifying the individual worker and channelling his energy into an individual project to raise standards, rather than into collective manifestations of displeasure. It killed two birds with one stone: revolution was postponed, and the working class, with its dreams of social elevation, made itself useful through the industrious development of its abilities.

The most interesting difference between the West-European and North American bourgeoisie on the one hand and the Russian bourgeoisie, which had attempted to introduce large-scale industrialisation of the Tsarist empire in the middle of the 19th century, on the other was that the European and American industrialists had the sense to use correctly the available tools and yardsticks and fix wages at a suitable level, whereas in Russia these tools were wholly absent, since there had only ever been minimal contact between the different classes. It was only in the polarised social atmosphere of Russia – where the dominant class was entirely isolated from and dismissive of (at least in terms of recognition) the demands of the working class – that it was possible to conduct a revolution.

The further an industrial society developed from feudal structures, the more flexible and developed its brand of capitalism became, and the smaller the risk for a workers' revolution. It was no coincidence that only the most feudal and agrarian societies in the developed world, Russia and China, suffered turbulent revolutions when industrialism began to accelerate. The difference between revisionist and revolutionary political development within the capitalist paradigm was directly related to the dominant class's access to information about the demands and wishes of the underclass. Violent revolution was to be avoided at all costs; the march of democratic socialism towards power was directly connected to the sophisticated mechanisms of capitalist society in this respect. Nominal power over the state became the basis for the compromise between the demands of the working class and the interests of the elite. Socialist revisionism is the ideology behind this ingenious compromise. When the majority is in charge of the state, how could a revolution be called for in the name of the people?

The terror of political correctness today is an act of bitter revenge by the minority against this worship of the majority. So-called weak groups gather in noisy alliances and demand rights in the form of quotas and special privileges. Pressure is exerted primarily through the media, and minorities with greater media power than others succeed better in this symbolic struggle for control of definitions. The result is the total impoverishment of political culture: the political arena is gradually stripped of substance and becomes the theatre for a frenzied battle between special interest groups. Opinion-based representation – the system upon which western democracy is based and according to which a spokesman represents his voters through the power of his opinions and not on the basis of gender or any other characteristic – is being replaced with a bizarre accounting exercise: every other one must be female, every fifth a pensioner, every tenth an immigrant, etc., etc. in absurdum. This spectacle is comprehensible in late capitalist society where perceptions of political power and the key roll of the mass-media still exist as a form of special-interest wishful-thinking. But in an informationalist paradigm it all looks like very poor theatre, and, if the consumtariat of the future is to have any pretensions to power against the wishes of the netocracy, it will have to find completely new ways of doing so.

On a purely material level, everything suggests that the underclass can continue to expect certain improvements; the social elevation of the underclass which began with capitalism will continue and assume new forms. But because the new underclass is characterised by its patterns of consumption and not by its relatively high living-standards, it is not possible to speak of any genuine reduction in the distance between the classes. A member of the consumtariat will not become a netocrat simply because he gets a larger apartment or a bigger car; he will be just as powerless as before, it is just that the price for his co-operation will have been corrected upwards.

When the supply of workers with a specific skill decreases, wages rise in the sector in question. If this increase in wages is not accompanied by a corresponding rise in productivity, the inevitable result is rising inflation. This is not in the interest of any of the parties within the market, which is why inflation is fiercely combated. The traditional method of raising production without ending up in an inflationary spiral is to promote population growth. A constantly growing population has always satisfied the constant demands of industry for a larger workforce. This sort of development cannot continue for ever; the levels of education which were necessary to increase competence, and the improvements in welfare which were necessary to maintain social stability generally lead to lower birth-rates. When birth-rates in the western world began to decline in the post-war period, there was a need for an alternative method of population growth to make up for the missing infants: large-scale immigration.

A temporary slump in the global economy during the oil crisis of the 1970s led to slower rates of growth and increasing unemployment in Western Europe and North America. Increased competition in the labour market meant that the previously popular immigrants became noticeably less popular, and the importation of labour was drastically reduced. But the dramatic increases in productivity which took place in the early 1990s with the beginnings of the IT-revolution have necessitated a more lenient attitude towards immigration. This is for the simple reason that the western economies are no longer self-sufficient in terms of labour. All over the industrialised world demographic development is following the same trend: there are more old people, and fewer young people.

The number of native workers is decreasing at the same time as the economy is in a period of rapid growth. The demand for external labour is increasing dramatically, which is why the western world is not only able to open its borders, but is actually forced to do so. As a result, there will be a lot of grand rhetoric from officials about multi-cultural society; the elite will combat isolationism and romanticised nationalism as hard as it can. Tolerance towards and curiosity about everything unknown will become an increasingly prominent characteristic of this overheated rhetoric. Naturally, this does not mean that purely ethnic and cultural conflicts, or more generally class-motivated conflicts will disappear. On the contrary, everything points towards heightened polarisation within western culture: the fear of widespread disturbances will become more justified.

It must be admitted that the new dominant class is genuinely cosmopolitan in outlook. The netocratic globalisation project is creating an electronic global culture. What this means in practice, however, is that netocrats in every country will unite on the basis of close contact and common interests, but without any tangible solidarity towards the immigrants who are mowing their lawns and driving underground trains. The netocrats will be defined by the fact that they manipulate information faster than they manage property or produce goods; their activity is thus linked to global networks, which means that their loyalties are virtually rather than regionally based. For them, multi-culturalism at home is partly a question of getting simple tasks performed, and partly a touch of exotic spice to life: an exciting variety of restaurants, clothes and entertainment. The netocrats will pay whatever is necessary to get their lawns cut and to buy tandoori chicken, but will not assume any additional obligations.

The new elite, in contrast to the old, does not perceive itself to have much to do with society in general. Thanks to new technology, it has the best possible means of avoiding troublesome taxation, but in return does not burden the welfare state to any great extent. Private insurance takes care of any private medical care that may be needed, private schools educate their children, privately employed guards will keep thieves and vandals away from their private property. The political establishment will become increasingly powerless, and the common sphere of society will diminish. In conjunction with this, both duties and rights will disappear.

The ideology which makes this state of things appear to be "natural" is meritocracy, once it has been fully implemented: nothing is determined in advance, neither provenance nor money will determine your fate, only your talent and industry. The same old underclass dream of a glittering social ascent, in other words, but the difference this time is that the possibilities of climbing up through the hierarchy are to a large extent real. Whether or not this is perceived as a good thing depends on the perspective you adopt. If by increased equality you mean an individual's improved possibilities to affect his or her own success, then equality will improve. But at the same time individual responsibility will increase, along with individual liability; personal failure will become much more personal. From a class-perspective meritocracy means, as the historian Christopher Lasch has pointed out, that the underclass is continually drained of talent and therefore also of prospective leaders. The elite, on the other hand, is strengthened by this constant circulation and the addition of new talent. Privilege becomes easier to legitimise if it is based on merit, because it is earned, at least to an extent, rather than inherited.

The new immigrants will largely be take their place in the subordinate underclass of the western world. On the other hand, their circumstances will be more or less bearable because their labour is genuinely needed and because growth sectors in their native countries will provide competitive alternatives. But differences in power, status and living standards will still be unavoidable. There are no signs that the specific religious and cultural identities of the various groups of immigrants will dissolve and melt together as a result of globalisation and migration; on the contrary, people without power and status build up their identities around their defining characteristics. Inverted racism is one possible scenario: violence will no longer be the preserve of badly-off natives and directed against welfare-sponging immigrants, but of badly-off immigrants against relatively well-off natives, or against other immigrant groups who are perceived as being more successful.

The growth of informational society will bring with it comprehensive migration. As far as the underclass is concerned, this will be a case, naturally enough, of moving from places with low rates of growth and relatively high birth-rates to areas where the reverse is the case. In North America this will lead to large-scale migration from south to north, and in Europe from east to west. But it will be the netocracy who will lead developments and decide their direction. The new elite is highly mobile, and will move for mainly cultural reasons to those places which are most attractive. This is principally a question of netocratic lifestyle migration. It will not matter how beneficial economic circumstances are: The cities and regions in question will lose out if they cannot offer a sufficiently enticing lifestyle and a sufficiently stimulating cultural environment. The consumtariat will have good reasons to adapt and migrate. It will be better to mow lawns, prepare tandoori chicken and collect their wages as citizens in areas of high demand and strong purchasing power.

In Europe it is already possible to see how the evolving netocracy is migrating towards a belt of large cities stretching from London in the north-west to Milan in the south-east. For the rest of the continent this means an growing and increasingly serious problem of depopulation, a so-called "brain drain" of the same sort as the migration from the European countryside to the cities during the 1900s, when talent and initiative were concentrated in economically dynamic urban areas. This urbanisation under capitalism is now being followed by nodalisation: extensive migration across national borders, from places on the cultural periphery to the cultural centres of the new paradigm, its geographical nodes or junctions. Only a few oases in the depopulated areas will have the foresight to exploit this development to their advantage, by recognising the implications of nodalisation in good time and drawing the correct conclusions as far as their own situation is concerned. The important thing is to create the preconditions for lifestyles which the netocracy finds attractive, of preparing fertile ground for stimulating cultural development. This process demands an unfailing understanding of what the netocracy finds desirable, which in turn will create a thriving market for a post-capitalist meta-netocracy, the lords' overlords.

One fundamental factor for success in this resuscitated system of medieval city states is that political responsibility will be delegated from the nation state to the cities themselves, and that regions rather than nations will be the primary unit of political structures. With globalisation the state will become a burden rather than an advantage; once matters of defence, foreign policy and monetary politics have been elevated to a supra-national level there will be no important matters left for national parliaments to discuss, while at the same time the globalisation project of the elite and the ghettoisation of the underclass will help to dissolve national identity. Dynamic cities which manage to escape enforced subsidisation of the countryside will be well positioned in this struggle. Like the medieval cities of the Hanseatic League before them, they will enter alliances with other cities when this is to their advantage, as it often will be.

It is all a matter of charming the netocracy, of playing upon its desires. The winner in this case really will take it all: wherever the netocracy goes, its servants will follow, and with a well-developed service sector the city in question will become even more attractive. Size is definitely not everything, because quantity is primarily a capitalist valuation. Even at the time of writing it is possible to see how the netocracy in the USA is finding its way to medium-sized cities like Seattle, Miami, Austin and San Francisco rather than to the mega-metropolises of New York and Los Angeles. The same thing is likely to happen in Europe and Asia. A careful balance of a wealth of different factors will matter more than size alone. This is a matter, naturally, of housing, infrastructure, communications, but these things alone are not enough. The netocrats are pack-animals, they seek out their like, and places where the range of lifestyles on offer is most varied. They will move wherever there is greatest cultural dynamism.

It is difficult to distinguish between cause and effect, because there is a constant interplay where the different levels mutually affect one another. Cultural climate affects migration and demographics, at the same time as these naturally affect the cultural climate. The fact that the population is gradually ageing means that guaranteed pensions will diminish in value, which in turn will mean that the age of retirement will begin to vary and will gradually creep upwards. The trend of the late 20th century towards youth culture – a sort of cultural puberty lasting well into adulthood – will become exacerbated by this, but even this development will not be unambiguous. The most striking pattern will be quite different: an increasingly wide gap between the culture of the netocratic elite and that of the passive consumtariat. In order to understand how this dynamic will function, it is necessary to look at how the media industry is developing.

The 20th century was a golden age for the mass-media. Technology made it possible first via radio, then via television, to reach out with the same message to an entire nation at the same time, then, via satellite, to the whole world. The ether-based media were the best propaganda instruments the world had ever seen. It is impossible to overestimate the importance of radio to the maintenance of national unity in Britain and the USA during the Second World War. Television played the leading role during the latter half of the century, and television's domination of mass culture has given considerable support to the slowly dying nation-state. It does not matter whether television was commercial, as in North America, or state-controlled, as in Europe. The central message was always the same: the nation is a "natural" entity, and beyond discussion, because nation and television audiences were one and the same thing. People watching the same programmes formed a connected and "naturally" segregated group. All of us television viewers must join together and behave like good citizens and consumers so that the wheels of the production apparatus turn smoothly.

Ironically it is the further development of the technology which has artificially kept the nation state and capitalism alive which is now burying the old paradigm. When the sitcom Cosby, in which all the central roles were played by black actors, became the most popular television programme in the USA during the 1980s, this was held up as a promising sign of the growing tolerance of the television media and its beneficial influence on its audience and on society as a whole. In actual fact this development was confirmation of a phenomenon that was already well-known within sociology: the fragmentation of the television audience and the gradual decline of the mass-media. The number of available channels increased but viewing figures fell. The decrease in television consumption (decreased consumption of each distributed unit) was a clear indication of the marginalisation of the media. "Broadcasting" was becoming more a matter of "narrowcasting"; instead of trying to capture large audiences, television channels were forced to concentrate on strictly limited segments of the audience.

The fact that Cosby topped the charts of viewing figures for a while was not primarily an indication of a new interest in racial issues and/or social justice within American television, but simply that unemployed, single black women had become the largest identifiable target group for television advertisers. This indicates not only a fragmentation of the audience and the media, but also an alarming brain drain. Nappies and washing powder are examples of products which are still worth advertising on television, whereas advertising trendy clothes or advanced electronics on television would be a waste of time and money. The new elite itself has little interest in consuming television, it is far too busy building networks with the help of new, interactive media. This has not prevented the netocrats from taking control of the medium of television and learning rapidly to use it to divert and anaesthetise the heterogeneous underclass which is united only in terms of its lowly status and increasing powerlessness.

At a stroke everything has become entertainment: the weather, the news, not to mention political journalism and election reports – all specially produced for an underclass of passive consumers, couch-potatoes with remote-controls sitting in the flickering light of the post-modern campfire, prepared to let themselves be entertained to sleep, with the chance of a win on some televised lottery as the highpoint of the week. For those who happen to wake up from their dormancy, television offers a studied and cynical level of pretended interaction. Ring and vote for the best player in the match or the best song in the programme, let us know what the subject for this evening's populist orchestrated debate should be! Naturally, all of this quasi-activity from the viewers is monitored in detail to help fine-tune the targeting of different groups.

The bourgeoisie has always had the greatest respect for television, and registers anxious delight at its infernal effectiveness as an instrument of propaganda, regarding it both as a wet-dream and as a terrible threat if, God forbid, it should ever get into the wrong hands. For the bourgeoisie television is the sexiest thing going, while the netocracy has a considerably more cynical view of it. Televisual entertainment still functions tolerably as opium for the masses, but its future is anything but glamorous. Television's fate, like all old media that technology has left behind, will be to provide content for the new interactive media, in the same way that the novel provided content for film, and film in turn provided content for television. This explains the netocracy's nonchalance towards television. The ceremonial pomp that once surrounded television is gone, the fortified studio bunkers and glittering office palaces are gone, grand, prestigious programming is a thing of the past, as is its continually increasing budget. Netocratic television is minimalist and functional, fluid and flexible, and has been out-sourced to a range of independent production companies. But all of this does not mean that television is not consciously stupid and stupefying; the netocracy will not channel its creativity into a medium whose future is behind it and whose audience it wants to control but not be part of.

In the final days of the capitalist paradigm it is still possible to regard television advertising as a necessary evil, an antidote to the unpalatable fact that someone has to pay for production, hopefully with something left over. In the world of netocratic television there will no longer be any real difference between adverts and programmes. Every detail will be product placement. The actors are products, selling themselves as products, when they are not occupied in selling actual products during commercial breaks. The products in turn are actors selling both themselves and the actors who appear in the adverts. The result is adverts for adverts for adverts, and, for the consumtariat who lack both the possibility and the capacity to participate actively, passive acceptance of the ordained rules of the game will be the only possible practical option. You pay for your entertainment with the minimum of attention, and exercise your participation the next time you choose between different brands of washing-powder. This is how you are told to realise yourself within the consumtariat and establish an individual lifestyle: by choosing washing-powder X or washing-powder Y for your dirty towels and underwear. Are you an environmentally-friendly user of X or a domestically economical user of Y? Choose your identity and win a free packet, here's the web-address!

Ever since informational society began to take shape in the 1970s philosophers and sociologists have questioned traditional concepts and distinctions such as work/leisure and production/consumption. To what extent have these conceptual pairs functioned as instruments of control under capitalism? How can we reasonably define different human activities in a social construct dictated by informationalism? Once again, we can see how old and tested concepts assume new meanings when their ecological context, and technology, changes. Consumption rather than production will be the activity which defines the underclass, where roughly equal amounts of resources will be expended whether one is employed or unemployed; the consumption of goods and services ought therefore, according to thinkers such as Baudrillard and Deleuze, to be regarded as an alternative form of production, one which is maintaining the machinery of society.

This revision of the definitions "consumption" and "production" is actually central to any understanding of the informationalist paradigm. According to capitalism an unemployed person, or anyone who stays at home from work during the day, has not achieved anything that day. Admittedly the day may have been filled with various practical tasks and social contacts, as well as a certain level of consumption, but none of this counts for anything in a paradigm where the only thing that counts is the production of goods and services, along with the surplus value this is expected to generate.

Production is productive and is therefore by definition something positive for a capitalist, whereas consumption is regarded as a negative, a diminution of a constructed value, a form of impious indulgence that one might grant oneself if one has been particularly productive. From a mobilistic point of view this division, the whole of this mechanistic chain of cause and effect, is purely illusory. In actual fact each activity presupposes the other; one is not possible without the other, they are two aspects of one and the same process. The consumer's desire for all the products and services that he or she cannot actually be said to need in any real sense is the decisive factor in the whole construction, and this desire has to be nurtured. The process is complicated, but the formula is simple: adverts + consumtariat = desire. The whole thing is an informationalist cycle, analogous to photosynthesis. Adverts are the sunlight, the consumtariat the diverse vegetation which transforms light into the energy which is the precondition of biology. This is the role of the consumtariat in the whole process: subordinate, but at the same time indispensable. What line of work the individual consumer is occupied with, if he or she works at all, is actually fairly irrelevant in this context.

We cannot determine if it is desire which produces goods and services, or if it is the goods and services which produce desire. The truth is that they produce and are produced by each other. It is pointless to try to distinguish one from the other when consumers are increasingly being paid to receive and react to adverts, and when consumers are increasingly using their attention as payment instead of money. Who is performing the real work, and who is paying whom for what? What seems at first to be trivial wordplay is actually the decisive factor in the struggle for power. In the capitalist paradigm the superior position of the bourgeoisie was based upon its power to define the work of the working class; under the new paradigm the new dominant class governs the new underclass by manipulating what could be called the consumption tasks of the consumtariat – more simply called desire. The fundamental difference between the netocracy and the consumtariat is thus that the former controls its own production of desire, whereas the latter obeys the orders of the former. Hence there is a vital symbolic value for netocracy in continually signifying in one's choice of lifestyle that one is independent of the consumptive production of manipulated desire, and thereby indicating one's social distance from the vulgar masses.

A netocratic lifestyle demands unique abilities and a particular overview. Because the moment that a product, a service or an idea becomes part of the advertised message, it is profaned, and destined for shabby mass-consumption.

What differentiates the netocracy is instead "imploitative" consumption: knowingly exclusive and minimalist, utterly free from directives. Netocrats travel to places without a tourist industry, listen to music that is not available from any record company, get their entertainment from subscription channels or net-sites which neither carry adverts nor advertise their own existence, consume goods and services that are never mentioned in the media and which are therefore unknown to the masses. This lifestyle can never be fixed, it will always be in a process of constant change. When the netocrats tire of one desire and the experience has lost its value, they can always throw it to the masses, recreating it for the consumtariat with the help of adverts, which also has its economic advantages. But whatever is reserved for the time being for the netocracy will always be unknown, incomprehensible and out of reach for the consumtariat.

In an age where automated factories or workers in low-cost regions of distant continents are increasingly responsible for the production of goods and services, work itself can no longer be the organising principle of society. The tiresome discussions of "the new economy" have to a lamentably large extent concentrated on glamorised slogans, capitalist-tinted descriptions of the future of the Internet, which suggest that this is a medium primarily intended for electronic mail-order sales. This suggests that nothing has fundamentally changed, as if new technology was merely a collection of trendy appliances with which we can happily go on mending and patching up old systems. The interesting thing about everything new is recognised all too seldom: that old definitions and concepts are being turned upside down, like for instance the production/consumption constellation, which forces us to revise all the associations we have regarding these concepts. A new paradigm implies new rules and new impulses in the struggle between the dominant class and the working class. Both capitalists and labourers worked, but it was the capitalists who dictated the terms. Both netocrats and the consumtariat consume, but once again it is the elite which is dictating the terms.

7.
The new biology and netocratic ethics

One important reason why "the new economy" seems so mysterious is that people generally, and economists in particular, have had such vague notions about how "the old economy" worked. Political ideologies and the social sciences that developed when the universities were granted a central position close to power in the capitalist paradigm obstinately followed the totalistic path, as we have already seen. National economists created sophisticated models that were impressive from every respect apart from the decisive fact that they gave a misrepresentative and useless image of economic reality. Their very starting-point was erroneous. Whether they represented the right or left of the political spectrum, the overriding goal was always the same: to build an all-encompassing theory that reduced the economy to an admittedly complicated, but coherent and manageable zero-sum game. Someone wins and someone loses, one presupposes the other, and all that was needed to avoid unnecessary friction and/or social injustice was distribution and regulation. Wishful-thinking about balance and order was mistakenly applied to a system whose normal state is characterised by constant change and a sizeable measure of destruction and annihilation.

Under feudalism the economy, or even the handling of money, were not things that serious people participated in. The production of goods and trade were consequently not regarded as sufficiently elevated or interesting to warrant philosophical analysis. Peasants grew crops and nurtured livestock, tradesmen engaged in bartering and aristocrats exacted tax in the form of goods and services. But when industrialisation, mechanisation and the transition from exchange of goods to coins and notes had reached a critical level, they brought with them dramatic social changes that gave rise to an entirely new set of questions. For instance, should the rapidly growing population be fed with the help of imported agricultural goods, or was it better to be self-sufficient in foodstuffs? Was it wise to protect your own country's peasants with the help of import tariffs? The rate of change was unparalleled, and the whole of the old rulebook was quickly outdated. So the first economic theory, the creed of wealth, was born, with the task of returning society to the harmonious social order which was supposed to have been lost.

The pattern for thought within the social sciences during the late 1700s and early 1800s was provided by the natural sciences in general, and physics in particular. The prestige of Newtonian physics was enormous, its achievements grandiose: uncovering the regulated mechanisms that governed the mysterious movements of the celestial bodies. What was required was an economic Newton, a philosopher who could 'discover' a law of gravity for economics, and uncover the eternal principles governing the divine order that was assumed to exist behind the apparent chaos. This was an impossible task, as economist Michael Rothschild and others have noted, for the simple reason that Newtonian physics was an unusable model, since it lacked a historical dimension. Time, or, to put it another way, a direction for physical processes, did not make its appearance in the natural sciences until the advent of thermodynamics later in the 1800s. For Newton, the universe is an unchanging perpetual motion machine, a sort of cosmic clock whose regulated movements never vary. This eternal, mechanical repetition was the whole point with Newton. His theory does not allow for qualitative change, and every economic model constructed with Newton as its pattern is consequently concerned with finding excuses for changes instead of understanding them. The goal was to achieve balance in the system, using suitable means, which is why the change that lies in the nature of things was regarded as unpleasant disruption.

Science finds what it seeks. The Scottish professor of philosophy Adam Smith, the central figure of classical national economics, consequently found an economic law of gravity: self-interest. When every individual is concerned with his own self-interest, the result, paradoxically, is the optimal state for social economics, according to Smith. Different people obviously have different talents and abilities, and it is when they have complete freedom to develop their talents and offer their services that the economy is maximised, to the good of everyone. Therefore regulations and import tariffs were bad, because the system was self-regulating. Expanding markets meant increased productivity, ran the optimistic gospel of laissez-faire liberalism. But even Smith did not accommodate change within the equation. His theory describes an imagined state of equilibrium, a liberal utopia, rather than the turbulence of reality. The system would automatically achieve balance and manage all possible disturbances internally if only it was left alone. But the idea that the system itself might undergo any decisive changes formed no part of Smith's calculations. That was a thought that could not be contemplated unless the illustrious Newtonian structure was abandoned, which it itself was unthinkable.

Smith's view of the economy as a well-oiled machine attracted many eager disciples. But not his optimism. David Ricardo worked from the theory that the amount of resources and goods in circulation on the market was finite. When the population is expanding and the number of consumers increases, increased demand leads to higher prices, especially of food. Hence equilibrium is disturbed in favour of wealthy land-owners; their profit corresponds to the consumers' and employees' loss in a desperate zero-sum game. There is no alternative way of keeping profits up other than keeping wages down, so strong social tension was to be expected. Fascination for the new economic science, thanks to Ricardo's theses, grew so large that national economics was believed to be capable of explaining everything. To a large extent it replaced philosophy, which was in a state of paralysis, as the metascience, the explanatory model for everything.

Ricardo's system was the theory used and incorporated by Karl Marx in his historical philosophy: economic and social oppositions, with their basis in the fact that the masses were being relentlessly sucked dry by the ruling class, escalate to the critical point where it becomes necessary at all costs to carry out a radical transformation of society, followed by a planned economy. For Marx, the economy was also a machine, but not well-oiled and self-generating as it was for Adam Smith, but a machine in real need of constant supervision and ideological direction. The political goal was a static economy, a regularly ticking clock. This is also in all essential respects the view of national economics that has been dominant during its two-hundred-year history, which has left concrete evidence in the form of various political programmes for regulation of the markets and for distribution. Buttons and levers have been pushed and pulled in the vain hope of achieving permanent stability in the system.

The various political camps have only differed from each other in emphasis. For liberals and conservatives, the right to private ownership has been sacred. The right's alternative to the monopoly scare of the left has been competition, but the fundamental problematics have remained the same: limited resources and an expanding population. Of course it is possible to increase productivity in, for instance, agriculture, but not in proportion to the increased manpower required, according to the law of diminishing returns which John Stuart Mill, one of the central figures of liberalism, formulated at about the same time as Marx and Engels were writing The Communist Manifesto. Increased welfare would never increase enough to satisfy the growing population. The best that could be hoped for was that a socio-economic status quo would be achieved, by appealing to the people's better nature and keeping the birth-rate down that way.

These ideas of the economy as a zero-sum game, and the curse of population growth came mostly from the cleric, economist and historian Thomas Malthus, whose influence on thought during the 1800s and early 1900s can scarcely be exaggerated. It was the fear of the terrible consequences of over-population that Malthus managed to inspire in his contemporaries which eventually led to birth-control measures being made available for a larger market, even if Malthus himself preached abstinence, and marriage later in life. His theory was based upon the necessary balance between the number of individuals and the amount of resources. According to Malthus's merciless principle, the population increases at a dramatically higher rate ("geometrically" or "exponentially") than the production of food does ("arithmetically"). This imbalance is unsustainable in the long run, and the excess must be removed one way or another. In nature this was regulated by famine and other catastrophes, and for human beings war was another option. Suffering and misery are inevitable anyway, and progress is a chimera. Helping the poor with handouts only makes things worse, because it will only lead to even more mouths to feed.

One of the greatest ironies in the history of ideas is that Malthus, the gloomy godfather of zero-sum philosophy, was also the great facilitator of evolution – in other words: the theory of change, development and process. The cul-de-sac which Malthus manoeuvred into made a whole new way of thinking possible. What Charles Darwin did was to apply history to history. He showed that change is not a deviation from any divine order, or a disturbance of any equilibrium in nature, but that constant change is itself the natural state. The species which had hitherto been regarded as eternal and constant, like perfect geometric figures, are actually historical products of other, extinct species. They are developed and adapt themselves according to circumstances. The problem for Darwin was that for a long time he lacked a driving-force in the process, an idea of the reasons for and mechanisms of change. He was clear about the idea of evolution, which was itself revolutionary, but did not understand how evolution itself actually functioned. Every peasant already knew that it was possible to evolve, to "improve" livestock and crops. It was a question of using specimens with the necessary characteristics for breeding and cultivation. But who was in charge of breeding and cultivation in nature? Who was responsible for the selection, and how did it take place? This was the big question.

To begin with, Darwin imagined that the same laws applied to species as to individuals; that they are born, mature, and die of biological necessity. After spending a year and several months getting lost in hopeless dead-ends, almost by chance he happened to read Malthus's theory of population. "For entertainment", he wrote in his diary. Suddenly everything fell into place. It was nature itself that oversaw selection. If more individuals are born into a population than the available food can sustain, the consequence is that many of these individuals die prematurely, without having had time to breed. Those who survive anyway, and, in the next step, go on to breed, are those individuals who are best suited to circumstances. The same process repeats itself for generation after generation. The cumulative result of this process is, eventually, the evolution of the species: natural selection rewards only very few of all possible variants. At the same time, the surrounding environment is continually changing, partly as a result of geological and climatic factors, partly as a result of its own internal dynamic. The altered species influence their own and others' circumstances, which in turn fuels further changes within the various species. There is no natural balance. The process never ends.

The ironic thing is that Darwin completely misunderstood Malthus. Or, if you want to take a more generous view of the matter: he made a breath-takingly original and ingenious interpretation of Malthus's utterly pessimistic reasoning, so that it fitted in better with his own thoughts. The struggle for survival which, for Malthus, was the root of all the world's ills, was for Darwin the mechanism which gave evolution a power and a direction towards increasingly sophisticated organisms. Thus biology took a decisive influence from economic philosophy, while economic thinking rejected biology and its historical perspective in favour of Newtonian physics and its static world-view. The same thing applied generally to sociology and the other new social sciences; Newton's cosmic perpetual motion machine formed the basic pattern for the creation of models under capitalism. The study of society took as its starting point a fictitious system in an artificial state of rest, and change was regarded as a disruptive anomaly. Thought was trapped within fixed, totalistic structures.

When Newton presented his theory he was hailed as the leading light of his age. When Darwin presented his theory he was practically treated as a criminal, and the theory of evolution has remained, outside the world of the natural sciences, extremely controversial for a remarkable length of time. Resistance to the theory of evolution was emotional rather than intellectual. The eternal and predictable, with its roots in Judeo-Christian religion and totalistic philosophy, always appealed to self-indulgent western dreams of Man's control and omnipotence. So change and coincidence were regarded with terror. This is why classical Newtonian physics has remained the model for natural sciences and continued to provide the model for the general world-view during the whole of the capitalist paradigm, even after physics itself had moved on from Newton, incorporated a historical dimension (with the formulation of the laws of thermodynamics), and become programmatically unpredictable, virtual, and generally exotic (thanks to quantum mechanics). We can therefore draw the conclusion that the old economics, like the old sociology and everything else scientifically "old", has been old for a very long time.

To a large degree, Darwin turned everything regarded as sacred on its head. The beautiful tableau of nature is not complete, but a permanent work in progress, and the question is whether it is actually particularly beautiful. The species alive now are neither original nor constant, and merely constitute a phase of the long development from simple to more complex organisms. The infinite wealth of variety and complexity in nature presupposes no divine creator, or any hidden intelligence of any kind, not even a plan; all that is needed are oceans of time. Evolution is a sort of algorithm, a numerical operation of immense scale applied to real life. Or a computer programme, if you like. Its function is to sift out the losers.

The American philosopher Daniel C Dennett has compared this process to a tennis tournament: two players meet, one survives and goes on to the next round, while the other is lost to oblivion. All knockout championships produce one winner, who has the qualities most favoured by the rules. If we are talking about tennis, then skill is largely decisive, but coin-tossing is purely about luck, or, to put it another way, the ability to avoid bad luck. It is obviously extremely unlikely that anyone would win at coin-tossing twenty times in a row, but if we organise a coin-tossing tournament with 1,048,575 participants, then there will certainly be someone who manages this. This sort of algorithm also does what it is supposed to: they unfailingly pick out a winner irrespective of how large the number of participants. Evolution is a form of knockout tournament whose rules are not only extremely complicated and full of previously unpredicted elements of chance; they are also changing the whole time. One round of coin-tossing, the next round a backwards, blindfolded, slalom sack-race. There will always be someone who wins and many, many losers. We are all winners: we who are writing this, you who are reading it, your friends and pets and houseplants, the trees in the woods and the worms in the soil, everything that is alive here and now. The losers are all the others. 99.99 per cent of all the species that have ever existed are now extinct. They were, quite simply, knocked out of the tournament.

We use a parallel logic when we pose classic hypothetical questions about how our lives would have looked if one event or other had never taken place, or if we had taken important decisions differently. What is this but a meme-Darwinian equivalent to Dennett's gene-Darwinian analysis? The other selves that we imagine we might have become instead of our current self, if the course of events had been different, could be said to be examples of inferior meme-Darwinian mutations, compared to the "I" who "survived", and who therefore enjoys the advantage of posing the hypothetical reasoning at the cost of "the extinct selves". This is the basis of the Foucauldian process of subjectification which is replacing individualism in netocratic society, and we shall return to this in the next chapter.

Nature is no picnic. Ruthless pruning promotes what is functional under the existing circumstances. Even our aesthetic comprehension of the existence which has survived with us, nature itself, is based upon the ingrained survival strategy of our genes. The beauty that we believe that we see in the colours of an orchid or the gaudy feathers of a peacock, and the fascination we feel at the giraffe's long neck, only become "beauty" and "fascination" to us because they confirm and underline evolutionary adaptability in our own great brain. Aesthetics are also a built-in genetic warning lamp. We appreciate a small child's attempts to walk and speak, or a dog's loyalty towards his master or mistress; the child and the dog are both useful and pleasurable to us, the usefulness and pleasure are mutual, the child and the dog appear to be aesthetically attractive, and we seem the same to them. At the same time we retreat from poisonous rattlesnakes and dustbins reeking of bacteria, because these phenomena are a threat to our own survival and have therefore been programmed into our genes as aesthetically repulsive.

Darwin's theory was far from watertight. One significant gap was the absence of a satisfactory explanation of how the winning characteristics were inherited from one generation to the next. A child generally shows a clear resemblance to its parents, but even children are not an even mixture of their parents' characteristics. One white and one black cat do not get a uniformly grey litter of kittens. If the parents' disposition to certain characteristics was completely mixed, the result would be an smoothing-out of all spectacular distinctive features, an even and uniform mass of bio-matter. But instead nature exhibits a constantly increasing level of complexity, variation and specialisation. How can this multiplicity and colourful display be explained?

The answer was given in a series of ground-breaking scientific discoveries in the latter half of the 1900s. In 1953 the researchers Francis Crick and James Watson described the unique structure of the DNA-molecule for the first time. By the beginning of the 1960s it was possible to read individual words written in genetic code, and by the middle of the same decade the whole of the code had been cracked, and now, at the beginning of the twenty-first century, the whole of the human genome has been mapped. The whole of our biological past, all the genetic preconditions for our future will be an open book. Quite literally. All the words in the book are written with the four chemical letters: A, C, G and T (adenine, cytosine, guanine and thiamine) in various combinations. Every living organism that has ever existed is created according to similar, now easily-read, instructions, all of them written in the same language.

The new genetics is one of the greatest intellectual revolutions ever. Suddenly biology is entirely digital. Life itself, the reproduction of cells and their creation of ordered systems of varying complexity, is a process which basically stems from information management. Life is fundamentally a question of the dispersal of information. Our genetic constitution is a collection of recipes, or programmes, for the production of proteins, which in turn regulate the body's chemistry. The body is the vessel of tissue which biological information has chosen to use. Thanks to pre-programmed information, the cells of the body know where they are and what they are supposed to do. No-one has to teach the egg how to become a chicken, the egg already knows. So the ancient information theorists were right in principle. It is information that breathes life into matter. One of the Nobel Prize-winners for medicine in 1969, Max Delbruck, suggested, possibly as a joke, that Aristotle ought to be awarded the same prize posthumously for the discovery of DNA. The old philosopher was right in that the form of the hen is already innate within the egg.

But genetic information does not merely constitute a recipe for anatomy, but also for behaviour. Classical humanists do not want to believe this, and insist that man has a soul, and that this soul and all spiritual phenomena are in some way independent of both the body and biology. Only soulless animals have instincts, we humans are above such things. Committed behaviourists are also sceptical, they claim that reflexes and behaviour are learned. Nature, and society, are regarded as one vast educational establishment, a sophisticated system of punishment and reward which forms the individual by encouraging certain behaviours and suppressing others. Man is, according to this view, an unwritten page. But the new genetics has smashed this way of seeing things. We can never learn things which we cannot learn. We will never be able to learn things that we lack the genetic predisposition for, however much we are rewarded or punished.

The brain is pre-programmed to be able to handle certain determined types of problem with the help of certain determined processes. The acquisition of language is one obvious example: our ability to understand and use grammar is innate, and research has localised one of the genes that is central in this context in chromosome number seven. A deviant "spelling" of this gene, what you might call a spelling mistake, means markedly lower linguistic capabilities (SLI, Specific Language Impairment) in otherwise entirely normally intelligent people: they lack the capacity to internalise grammatical structures. This means that every new word they encounter really is new to them; for every verb, for instance, they must learn each conjugated form separately, along with every plural form of each new noun, and so on. Quite simply, the pattern is not instinctively clear to them.

This means it is not possible to replace instincts with learning. Of course, someone with SLI can learn to communicate with the outside world, albeit with certain difficulties in understanding and making him- or herself understood, but he or she can never learn to think grammatically. That we humans, in contrast to our close relatives among the apes, have learned to use grammatically constructed language is not because we have been more industrious and have tried harder than they have, but that, thanks to genetic changes, we have developed new, species-specific instincts. Language – which in its spoken and, later, written forms has been the dominant means of cultural transfer and development – has, without any doubt, its roots in biology. We have learned what we have been able to learn. On a fundamental level, it is the information in our genes that determines "who we are".

The consequence of this is that there really is a "human nature", and that this influences to a high degree not just our capacity for language acquisition, but our behaviour and our culture generally. A new-born child is not a blank page, but the carrier of a programme that admittedly allows for an enormous amount of development, learning and interactivity with the surrounding world, but which, in spite of this, has its special structures and its special limitations which are ultimately determined by biological history. The brain is a product of evolution, and from this follows a whole succession of collective, fundamental thought-patterns. Genes keep culture on a leash, as Edward O Wilson suggested.

But the idea that there is a connection between biology and society is still met with bitter opposition from many directions. The same ideas are often summed up, in a semantically dubious way, as "reductionism" or "determinism". This opposition is largely politically motivated: one basic thesis in Marxism is that society entirely shapes the citizen's consciousness, and that a new society would mean the creation of a completely new person. If it turns out that biology is the ultimate determinant, then the Marxist left will have to think again. But even within the social and human sciences, the accepted stand-point of the entire twentieth century was that biological evolution and cultural development are two separate phenomena with no points of contact. What has been of interest for research has consequently been questions about how the social environment has shaped human behaviour, rather than how man's social instincts have shaped society.

Even this point of view has its ideological causes, mainly the fact that coarse misinterpretations of the theory of evolution, together with other quasi-sciences, have been used by a long succession of charlatans to legitimise various racist and other suspect ideologies. "Lower standing cultures" have been "explained" with reference to a supposed innate refinement, and so on. The desire to dissociate from such "vulgar biologisms" is in itself quite understandable, but any consequent attitude of "guilt by association" constitutes just as destructive an act of intellectually blinkered thinking in the other direction. Blinkers are always blinkers, no matter how noble the reason for wearing them.

The fundamental idea of the theory of evolution, that it is chance which is decisive, is the complete opposite of the starting-point of vulgar biologisms. All talk of different races and their varying genetic disposition for highly developed culture is complete nonsense. The fact that people in the fertile crescent of the Middle East abandoned life as hunter-gatherers at an early stage and built the first agricultural society was purely the result of the fact that circumstances there were right. The climate was favourable, and, above all, there were plants which were suited to domestication and cultivation on a large scale. One led to the other in an increasingly advanced feedback loop. Increased access to food and fixed dwellings led to population growth, which in turn created dramatically improved conditions for increased specialisation and a more advanced social structure. Which in turn generated even more economic growth, and, later on, cathedrals, sonnets and string quartets.

But the right conditions alone are not sufficient as an explanation. Any friend of order might ask: the conditions for what? It is not possible to avoid biology and social instincts any more. The humanist bourgeoisie were passionate about the refinement of culture that they saw as characteristic of Man, and which raised him above beasts and the law of the jungle. Like a fundamentalist sect, humanism insisted on Man's unique position, sort of floating just above the rest of nature. That culture had its roots in biology, and was an indivisible part of it, was unthinkable. But what possible alternative is there? If the construction of society and culture do not have an evolutionary basis, what sort of origin might it have? The answer can be found among the absolutes of religion and myth: culture as miraculous creation, a gift to humanity from God knows who. For the netocracy which is now assuming power, these metaphysical bolt-holes lack intellectual credibility. So the wall between nature and culture is being torn down and humanism is going to its grave.

When we compare cultural evolution with biological, this is not only a matter of a spectacular metaphor. It is a question of a scientific and socio-philosophical earthquake. The Newtonian/static/mechanical view of society, culture and the economy is finally relinquishing is cast-iron grip on thought now that the capitalist paradigm is drawing to its close. Physics, in particular Newtonian physics, is no longer the model science. The twenty-first century belongs to biology. An entirely new world-view is taking shape before our eyes. We are talking about a world beyond humanism, trans-humanism.

Genetics has one important characteristic which makes it irresistibly interesting: it works. Advances within plant and animal production have been spectacular in recent years. It is no longer possible to question seriously either the methods or the theoretical basis. At the same time, our knowledge about the human species is increasing at a dizzying rate. The mapping of the human genome means the identification of the circa 100,000 genes which, spread over 23 chromosomes, make up the chemical formula for a human being. The whole of this incomprehensibly long text – about a billion words, which is the equivalent of 800 Bibles – will be readable, which will give us detailed information about both our past and our future. Moreover, the text can be edited.

With knowledge of his genetic predispositions, man can, for the first time in his history, plan his life from genuinely fundamental information: he can choose an education and a career that suits him, create children with a partner who possesses a complementary set of genes, choose not to eat harmful foodstuffs, and so on. Employers and authorities will have access to aptitude tests worth the name. The meritocracy, as Swedish biologist Thorbjörn Fagerström has pointed out, will materialise in the entirely new form of a "genocracy". This development will naturally be met with protests, not least from classical humanists who take offence at the fact that people's inherited aptitudes are compared and ranked. But it will be difficult to claim that the process is not "natural". Nothing could be more natural than comparisons and ranking, that is what natural selection is all about, and what principal of selection could be more "natural" than the genetic? This development will be unstoppable, for the simple reason that its application actually works, and that the principle of "right man, right job" is so valuable for the interested parties. It will be claimed that certain job categories are so important that for that particular case the end justifies the means. And that is where the dam will burst. When a taboo has been transgressed in one specific area, it is impossible to maintain this taboo for society in general. Particularly in a plurarchic society.

The connection between sexuality and reproduction is disappearing. Sex is becoming more of a hobby, an expression of identity, with neither desired nor undesired consequences. Instead, reproduction will be managed under orderly conditions in laboratories. Who is the parent of whom will be a complicated question when sex-cells, which in principle could come from anywhere, are installed in artificial wombs. "Pregnancies" will be carefully monitored. When it is time for the "birth", the chance of surprises will be drastically reduced. As a result of gene-manipulation cancer, Alzheimer's, allergies and a whole list of other illnesses will be preventable at the embryo stage. It will be possible, to a large extent, to shape, or rather programme, your "offspring". And even to add qualities that we hardly used to think of as "human".

This development is being hastened by the demise of the belief in a perfect "natural order" regulating how reproduction should happen. Means of reproduction have actually varied a lot during the evolutionary process. Our original forbears practised cell-budding. Later on they laid eggs and determined the sex of their offspring by regulating the temperature around the eggs. The reason why gender-determining genes got the upper-hand was that every individual, even at an early stage, needed to prepare for the gender-determined tasks that awaited after birth. So the only evolutionary "natural" is change itself. One humanistic argument against placing an artificially inseminated egg in an artificial womb is that this would be a typical example of interference in and assault on nature. The problem with this sort of reasoning is that it assumes that culture and nature are two essentially different phenomena in opposition to each other. But this schematic way of seeing things is completely outdated in the informationalist paradigm. Culture is a new version of nature: Nature 2.0.

The increasingly marginalised humanist institutions of power from capitalism will raise demands for restrictive laws governing genetic technology. They will demand observation and strict control by the state and academic experts of all research which comes into contact with the old taboos of the bourgeoisie. In many cases these demands will get a response from politicians, and in certain cases more or less extensive regulation of genetic experimentation has already come into force. But this is of little importance. The constantly weakening position of the nation-state in comparison to new, growing forces – such as the adventurous netocracy and the expansive multinational biotechnology companies – will impose insurmountable obstacles for these political efforts. The most advanced genetic research is already carried out in closed laboratories and under great secrecy on privately owned domains, which makes it extremely difficult to control. Besides, the West's strongly Judeo-Christian-coloured attitudes about the sanctity of the unique individual is anything but universal. In other parts of the world, in Asia for instance, there is a far less sentimental view of the matter, and research carries on unhindered.

We can count on a shaky and conflict-filled process of acclimatisation. Radical biomedicinal advances have traditionally aroused feverish debate and met resistance from groups which believe their moral authority to be threatened. Corneal transplants are an instructive example. When it became medically possible to save people's sight with the help of fresh corneas from the recently deceased, this met with strong resistance and the method was banned in Great Britain, amongst other countries. The method was classified as unethical on the basis of its use of body-parts from the deceased. Their dead bodies were regarded, ironically, both as sacred and impure at the same time. But today this method is a routine procedure. Saving the sight of the living became, as information about the procedure spread and superseded old moralisations, more important than maintaining the sanctity of the corpse.

In a society without a central moral authority, even without a parliament, it will be this silent battle for power between these interest groups which will determine what is regarded as acceptable and, above all, what is practised. Something which is unethical today might well be totally accepted tomorrow. In spite of everything, people will willingly accept the new medicines offered by applied genetic technology. They might suddenly be able to accept the specially created organs of manipulated pigs for transplantation, if they or someone close to them is in dire need of them. And when they have the chance to choose, people would like to have well-formed children, without a known predisposition to cancer, for instance. Pragmatism directs medical ethics, not vice versa. Netocratic ethics are therefore a hyper-biological pragmatism.

It is hard to imagine that people would choose to do without the very possibility of choosing. From a historical point of view, the choice of forbidding choice is only applicable within extremely hard-line religious sects (like for instance the Amish people's rejection of electricity). It is simply not in our genes, as the history of science clearly demonstrates. We are curious by nature, and extremely adaptable. On the basis of all available information, within the near-future it will be possible to create trans-genetic clones of ourselves, completely identical except in the aspects we choose to modify: not near-sighted, not bald, whatever we want. These lightly retouched copies could even be used as living stores of reserve parts; perhaps we will need, for instance, a fresh new liver to replace the old one that we sacrificed to drink?

The loss of any central political power makes this development entirely possible, even if a majority of citizens might be negative towards cloning, for instance. In informational society it is not the voter but the prosperous consumer who is in charge, a thought which has been proposed by the zoologist Matt Ridley. This was what happened with test-tube fertilisation. A sufficient number of childless couples showed themselves sufficiently keen and sufficiently wealthy, and the possibility was there. Today test-tube fertilisation is a routine procedure. Within the near future we will see numerous examples of this sort of "netocratic decision-making"; beyond the classical political model, and beyond the influence of the majority.

What is fundamental for all these rapid changes is that the concept of "natural" is completely losing its value-content. The more we learn about our biological history, and the more we learn about the history of culture and the construction of human society, the clearer it becomes that the oppositional relationship between these two phenomena, which has been held to be self-evident, is the only thing in this context which is genuinely artificial. Nature and culture are both fundamentally immensely complicated systems for the management of information. They both follow exactly the same law: natural selection. They both demonstrate the same inherent logic: a movement away from the simple and particular towards ever more sophisticated interaction on an increasingly large scale. In the new world-view which is rapidly taking shape – and here we are not talking about objective truths, it is the new paradigm which will ultimately determine what can be thought – nature and culture are two complementary sides of one and the same thing: evolution.

In the beginning the earth was an energy-filled soup where the simplest cells imaginable, the forefathers of our cells, drifted about and multiplied. The first step towards cathedrals and string quartets was taken when a number of these original cells bumped into what you might call a parasite, a bacterium, the forefather of our mitochondria (the organelle which manages the cells' metabolism). The meeting was not friendly: either the cell tried to swallow the parasite but failed to digest it, or the parasite tried to invade the cell but failed to kill it. Either way, the result was a collaboration which benefited both parties: a new type of cell where different elements managed different tasks. This cell, which practised internal division of labour, was the precondition for an even more developed biological collaboration in the form of multi-celled organisms where different cells were given different roles. Since then natural selection has, incredibly slowly, created ever more complicated forms of collaboration between cells by deselecting competitors that were less prepared to collaborate. The genes which have been suitable for advanced integration have been favoured.

Increased specialisation and co-ordination made considerable advances in productivity possible. Together, the cells became "intelligent". They constructed what the biologist Richard Dawkins has called "survival machines": gradually more refined animal bodies which were instructed to do intelligent things in certain situations, for instance to seek out warmer places when there was a threat of frost. But size is not everything, as every businessman knows. A large amount of co-ordination also entailed costs in the form of increased use of energy. Natural selection weighed one against the other. For this reason, bigger and bigger organisms were not the obvious solution to all problems in the harsh environment of the genes, which explains why co-operation between cells has assumed other, even more ingenious forms, with collaboration between individuals, schools, flocks and societies. A society which looks after its members' interests benefits the genes involved to the highest extent, increasing their possibility of surviving and reproducing.

The pattern in history is the same: one plus one equals more than two, co-operation benefits all parties involved. Natural selection prefers people and societies that learn to play non-zero-sum games (in contrast to zero-sum games) together. What happens when nomadic hunter/gatherer tribes settle down and begin to work the earth is that the conditions for a successful non-zero-sum game are radically improved. Over time a constructive spiral develops: technological and economic progress gives rise to population growth, which in turn means better conditions for further technological and economic progress. The towns which eventually develop are sufficiently tightly populated to support functioning markets, and economic growth gains still more speed. Contact between the different towns leads to the organisation of a more comprehensive system of co-operation. Thanks to ground-breaking technological breakthroughs, man is able to cross difficult thresholds and develop ever more advanced forms of non-zero-sum game.

But every force has a counter-force. History demonstrates an intricate dynamic between zero-sum games and non-zero-sum games. All the wars which have laid waste empires and cost countless human lives are excellent examples of explicit zero-sum games, or even minus-sum games. What someone wins is lost by someone else, at the same time as enormous resources go to waste. This does not stop the final number from actually being a positive. The threat of war unites a society and leads to the establishment of alliances with other societies, such as when the Greek city-states united to combat the Persians. The journalist Robert Wright, who wrote Non Zero, one of the books in which the biologically-influenced world view appears most clearly, suggested that war has a sort of coagulating effect by forcing people into organic solidarity; war provides an external threat which necessitates various forms of close co-operation. This is a thesis which can be seen as a parallel to the mobilistics idea that man's knowledge of his ultimate death (war against illness and ageing) has a central function in the creation of his individual identity.

The actual situation, at the transition between an old and a new paradigm, is ambiguous. On the one hand we can see in the collapse of the nation state a tribalisation: how larger entities are broken up into smaller ones where identity and loyalty are bound to different subcultures. On the other hand the declining nation state is being replaced by supra-state institutions, in politics, economics and culture. On the one hand fragmentation, on the other integration. Wright calls this phenomenon "fragmegration". But the new information technology which is driving development has its own programme, it offers co-operation and non-zero-sum games. The struggling indigenous population, fighting for increased rights and varying degrees of self-determination, co-operates with other groups in the same situation in world-wide electronic networks. Isolation is not a strategy with a future; the tension between local and global is only apparent in the virtual world. For Wright and others, the current situation is bringing to the fore the old question of a global state, a matter we will address in more detail in our next book.

Of course, what is relevant for the evolutionary development of society and culture is also valid for the economy, as Michael Rothschild insists. The market economy is "natural": an unplanned but still highly structured ecological system in a state of constant change. There is no equilibrium, no lasting state of repose. Even here the laws of natural selection apply, favouring the actors who are skilful non-zero-sum players with the capacity to build strategic alliances. Poorly organised companies that cannot learn new methods and therefore cannot cope with competition under the rules which apply within their particular niche of the market are sifted out, which gives more space for new players. Even conspicuous consumption of luxuries has its evolutionary logic: the sexiest people attract a partner most successfully, and sexiness in nature is often synonymous with big horns or colourful tail-feathers – an extravagant waste of resources, in other words. Just as rationality is not always most rational, so effectivity is not always most effective, either in nature or culture. The netocrats' imploitative consumption expresses the same thing, but in another way: consumption as a mark of status or seductive artistry. We are talking about an intuitively guided, trans-rational economy which plays upon the exhibition of exaggerated resources, an economy which would drive classically educated accountants and stock exchange analysts mad with its playfulness and transgressions against the laws of rationalism.

This means that the mythical concept "the new economy", like "globalisation", is actually two completely different things. To begin with, it is the old economy appearing in an entirely new light, as a result of old models and thought-processes being replaced for new ones as a consequence of the current paradigm shift. The new models which are constructed on the basis of these new insights are likely to be considerably more clarifying than the old ones, as far as both new and old are concerned. And secondly, the new information technology gives willing learners the chance to play entirely new non-zero-sum games with each other. Co-operation is seeking out unexpected paths, completely new categories of information are becoming valuable, strategic alliances are becoming increasingly extensive and transgressive of boundaries. Producers, suppliers, distributors and consumers are being bound ever tighter into digital networks. A genuine understanding of the former will dispel much of the confusion surrounding the latter. And power in informational society will end up with those who understand, and manage to dispel the confusion.

8.

THE CONVULSIONS OF COLLECTIVITY, THE DEATH OF MAN AND THE VIRTUAL SUBJECT

One consequence of the revolutionary advances within genetics, and biology's increasingly dominant position within our thought, is a total relativisation of the concept "individual". If by individual we mean the ultimate instance of control, literally indivisible, then the individual is looking more and more like a sheer illusion. An analysis of the interaction between body, brain and genes in different situations reveals that there is no instance of control. The genes release different chemical reactions which influence the body, but the genes, in turn, are activated by the brain, whose decisions are themselves instinctive reactions to external stimuli, via the body. We could say that it is the situation that decides, but since we ourselves are part of the situation, we would inevitably end up in a feedback loop without beginning or end. No-one decides. What we find in place of the individual that we thought existed, is a sort of turbulent market economy in micro-format, where a wealth of factors contribute; a constantly changing tension between different forces and counter-forces. The subject's undecided nature and mutability is leaving the philosophical sphere in netocratic society and becoming instead the acute and tangible stuff of the everyday for the common man.

One basic precondition for the possession of power is access to and control over information. In feudal society the flow of information was strictly controlled by those in power. The average individual's entire social interface with the outside world throughout their life was strictly limited, and was restricted in total to roughly as many people as could fit into in the local church or would turn up to village dances. Trade was limited, communications with other regions negligible, news from outside limited and strictly regulated. The representatives of the church and the aristocracy had power over a relatively comprehensive flow of information, which they passed on to the underclass in small and carefully weighed-out portions in the form of sermons and decrees, naturally adapted to serve the interests of the elite. The news that reached the village through legitimate channels therefore first passed through a strict control apparatus with branches in the monasteries and manor-houses. Everything possible was done to silence unauthorised sources of news and brand them as criminal; the many itinerant travellers who existed in spite of everything were regarded as bandits, a lawless rabble, and they retained that label in the history books of the capitalist paradigm. The enemy was by definition strange, and strangers were by definition the enemy.

The individual traveller was supposed to offer simple handicrafts and entertainment; he might make decorations, or perhaps music at village fairs. But he was not a trusted member of the village community. The wanderer had no rights, in principle, and was subordinate to the whims of the local aristocratic and religious leadership. Having a fixed abode was one absolute minimum requirement for anyone wanting to get married, the ritual by which society sanctioned reproduction. People outside the village community were worthless. For a soldier who was ordered into battle, the punishment for refusal to defend the local community against the evil barbarians was ex-communication, not just from the social community (the aristocracy's threat) but also from heavenly paradise (the church's threat).

Feudal structures began to give way when the flow of information could no longer be monitored effectively as a result of an increase in organised trade. The first tendencies towards capitalism emerged when communication between towns intensified, which meant that each individual's social interface expanded enormously. Tradesmen in towns around the Baltic Sea formed the Hanseatic League to protect their common interests and to encourage trade in any way possible, and this confederation became strong enough to set itself against the Danish king and keep trade routes free from piracy. The Italian city-states came together in the Lega Lombarda in order to form a united front against the Germano-Roman emperor's demands for obeisance. Co-operation was worthwhile. A new power structure was growing stronger.

Villages grew into towns. Urbanisation and the increasing power of the young bourgeoisie meant that information flourished and was distributed to a completely different extent to before, and it became necessary to adapt the use of power to these new circumstances. Feudal towns had been completely surrounded by walls and moats, partly because they needed physical protection from the outside world, but partly because this facilitated the operation of effective tollgates and thus a considerable income for the town. Capitalist towns, on the other hand, grew so quickly that this sort of enclosure and demarcation became unsustainable. It is actually questionable whether we can talk of towns in their real sense before early capitalism; even imperial Rome consisted largely of a loosely-connected community of villages compared to the town-development that arose with capitalism.

As a result of these changes, the town rulers could no longer exercise unlimited power over a clearly demarcated area. The arrival of capitalism is clearly visible geographically, when town walls stopped functioning as a barrier between town and country.

The aspirations to power of capitalist cities therefore extended far beyond the boundaries of the city itself, reaching out into the surrounding countryside and not stopping until it came up against a natural barrier in the form of mountains, seas or large rivers. The reason for this expansion was not merely that the populations of the cities was growing, but that this population was in need of regular deliveries of food and that the new factories needed regular supplies of raw materials in considerably greater quantities than the main occupations under previous conditions, concentrated on the market-square, had done. The city therefore colonised the surrounding area and within this naturally defined area a common identity developed, based upon appearance, language, mythology, articles of faith and customs. Thus was the modern nation created.

Thanks to improved communications and rapidly growing urban populations who lacked any restrictive regional loyalty, the new social identity could spread to cover a considerably larger area and many times more people than before. Expansion was a necessity, but must be kept within reasonable limits. Like all biological organisms, it was necessary for the nation to establish strict boundaries with the outside world. Self-protection requires the establishment of fixed boundaries, so that it is clear what is actually being defended, and so that limited resources are not wasted in feeding the rest of the world. Therefore it was in the nation's interest to mark clearly who "we" were, in contrast to "the others"; consequently power over information was used to this end. Censuses, registration and cataloguing of all settled inhabitants who could claim national identity and belonging were organised. The state accumulated more and more tasks, which necessitated increases in taxation since the growing administrative apparatus was swallowing ever more resources. As a result, the power of the state grew.

The increasingly powerful bourgeoisie of the rapidly growing cities laid claim to the surrounding countryside in order to secure their supplies and protect their power. This happened by the bourgeoisie using its new-found position of strength to compel or out-manoeuvre and marginalise the rulers of the countryside, the aristocracy. Regardless of the formal constitution, in reality the state assumed the right of the monarch and the aristocracy to levy tax by shifting the right to impose toll-charges from town-walls to national boundaries. The nation replaced the old town as the geographic basis of citizenship, which was confirmed by the supreme instance of capitalist identity: the passport.

In the new geographical entity, power was organised in a centralised system, which meant that all power came from, and all information was directed towards, a clearly defined centre: the capital. This organisation and the image of a centre surrounded by increasingly peripheral outposts left its mark on contemporary thinking about society and the world. The model was still the Christian heaven with God and his angels. The word "capitalism" has its origin, like the English word "capital" (city), in the Latin "caput" (head). The capital was thus the nation's head: the giver of orders and information centre, as well as the symbol of the values around which the nation was gathered.

This new age required a new individual, a new ideal of humanity adapted to the needs of the state and the market; the feudal peasant patiently ploughing his fields and waiting for the return of Christ was far too passive and intellectually sluggish, and therefore not sufficiently receptive to the sophisticated propaganda of the new age. In rapid succession a clutch of new concepts appeared, forming the basis for the definition of the new individual: the nation, race, citizenship, income tax, education, mental illness, criminality, the foreigner. Around all of this was formed the "common ground" which was the cohesive cement of the nation state. The bourgeoisie was protecting its newly-won monopoly on information as best it could.

According to this model, the citizen should not need to be threatened in order to make him protect his country. Instead, the sense of national belonging and the values it contained should be seen as so valuable that no-one would hesitate to take up arms whenever the nation's sovereignty was threatened from ill-intentioned neighbouring peoples. Consequently considerable energy was expended on the production of a nationalistic culture, through the mythologisation of the nation's origins and a romanticisation of its history. Poets conjured forth an heroic past. The traditions of the nation were supposed to be linked to its geographic area in a sacred symbiosis. Hence the myth of the origin of the nation arose.

But at the same time this required the demonisation of the outside world. The feelings towards other nationalities that were propagated were a mixture of fear and loathing. The very essence of nationalism is based upon establishing distance from, and a disdain for anything foreign and unknown. Confused race-biology was one completely logical consequence of this development, a frenetic desire for empirical support for feelings of superiority, the feeling that gilded nationalism and raised national citizenship to something elevated and sacred. Aside from this, racism had another attractive function. Nationalism alone could only summon forth suitable demons in time of war and conflict with neighbouring countries. But thanks to racism, demons could be produced even in peacetime in the form of internal minorities with deviant physical characteristics and cultural traditions, like for instance Jews or gypsies, and oppression of these groups could thus be legitimised. The nation state was thereby assured that there would always be demons to hand, and scapegoats on which to blame all shortcomings.

The fateful consequences of these centralised mechanisms of nationalism kick-started the accelerating decline of nationalism and the inevitable collapse that is evident in late-capitalist society. The eternal human search for a basic sense of belonging has, first through the appearance of popular culture in the mass media in the post-war period, then, above all, through the establishment of electronic tribes on the Internet, found credible alternatives to floundering nationalism. In informational society virtual subcultures are replacing feudalism's village communities and capitalism's national communities as the basis for man's social identity. In a society like this, obviously no-one is prepared to die for his country. National boundaries and their physical guards, military organisations, are imploding. New boundaries between social groups are being established with great seriousness in the virtual world.

It is therefore interesting to note that the most stiff-necked extreme nationalists – neo-Nazi groups in western Europe, fascists fighting for regional self-determination in eastern Europe, isolationist and historically romanticising fundamentalists in North America, east Asia and the Middle East – i.e.: the nationalists who for the sake of consistency are still clutching the banner of racism – are the only groups who have succeeded in building functional electronic networks which are based entirely upon national identity. This is a fragmegrational phenomenon (see previous chapter); strictly disciplined and organised subcultures promoting their interests by constructing strategic alliances with like-minded people on the net. All "milder" forms of nationalism – flags, traditions, pride – have, in contrast, never gained a foothold in the virtual world and therefore have no future in informational society. The boundaries of the nation state are today as irrelevant as moats were at the breakthrough of capitalism. Besides, the disintegration of the nation state is strengthened and accelerated when its institutions stand helpless in the face of the task of controlling and, above all, taxing "the new economy".

One problem which is becoming acute as a consequence of this development is the constant undermining of the authority of the nation state's institutions. When laws in key areas can no longer be maintained, this drastically reduces the citizens' respect for the legislative collective and the judicial apparatus; this is particularly the case among the groups who are being favoured economically and in terms of status by technological development. Those who are clinging most tightly to the wreckage of the nation state are in part members of the old, obsolete dominant class who can see their position and their privileges sinking to the bottom, and in part members of the new underclass who consciously or unconsciously realise that no changes in the current direction will do them any good.

For the netocrats, on the other hand, the nation state and its barriers appear mostly as an irrational, but passing, cause of irritation impeding the flow of traffic in the global village. The remnants of nationalism are, in the eyes of the netocrats, a shameful sickness that ought to have been eradicated by now, a sort of mental handicap which is maintaining the old dominant class in a state of impotence and decadence, and is suppressing the degraded underclass in a permanent state of inferiority. In short: an epidemic and a delusion which it is a humanitarian act of charity to combat. For the netocracy, the raising of the national flag is the most offensive example of vulgarity and bad taste. This of course does not prevent the new dominant class from exploiting every opportunity to project the symbols and trademarks that represent their own electronic sects. Symbols which, ironically enough, are often superannuated old national flags. The domain address of the Soviet Union, .su, was for instance one of the most popular Internet addresses among netocrats after the collapse of the Soviet Union.

The police and the judicial system are increasingly powerless in the face of growing electronic crime, a sort of global networking mafia where the motivation for criminal activity is more the creation of identity and increased status within their own group than economic profit. This is leading to shrill political cries for increased resources for the police and public prosecutors. But just as national tax authorities find themselves in a hopelessly inferior position compared to the mobile netocracy which is in charge of the boundary-busting virtual economy, so national criminal justice authorities are coming off badly against a "criminality" that has no specific geographical location.

At the same time the bourgeois family is rapidly being undermined: when industrial production falls dramatically, to the benefit of information management and an expanding service sector, the conditions of the labour market are altered. Poorly educated men are becoming superfluous whereas a wealth of new opportunities is opening up for women, which is in turn undermining the already highly-pressured nuclear family. The instance of divorce has risen constantly in the west since the beginning of the 1960s. These factors are working together to form a serious, destabilising crisis in informational society. Great amounts of uncontrolled energy are released when the social institutions of the old paradigm collapse and the very basis for the social identity of the majority of people disappears.

In parallel with the development of capital cities during early capitalism, another completely different type of city culture developed: what the American philosopher and historian Manuel De Landa calls a metropolis, a form of city that first became possible in capitalist society. The metropolis did not form the centre of the nation, either politically, culturally or geographically, unlike the capital; instead it was positioned on the coast, in connection with the increasingly well-trafficked and important intercontinental sea lanes, and was therefore in several important respects isolated from the rest of the nation. While the capital imported its workforce and raw materials from the surrounding countryside, the metropolis built its expansion mainly on international trade. The imported raw materials and tradable commodities came from foreign countries and, not least, from other metropolises. These independent seaports became junctions for vital transport routes, through which passed an enormous traffic of people, base products and luxury goods desired by the capital cities and their nations. The metropolises' power was not based upon control of territory, but on control of financial flow.

The capital and the metropolis had entirely different functions in capitalist society. While the capital embodied the structure of power itself, the metropolis was largely free of the restrictive laws and regulations that applied to the nation state. Therefore the metropolis became a base for various activities and phenomena which were regarded, for various reasons, by nationalism's propagandists and the rulers of the state as morally dubious and dangerous to society. It might be prostitution or the slave-trade, or the loaning of money at interest, something which was regarded as fundamentally suspicious during the late Middle Ages. For this reason the metropolis became an arena for all sorts of experimentation in both lifestyles and thinking, because it provided a level of freedom that was unthinkable within the nation state itself. In the metropolis the feared and hated nomad could find a haven. People came and went, this was not only accepted but entirely in line with the idea of the metropolis; less attention was paid to control, and nomadic lifestyles developed over time to the point where they were almost the stylistic ideal of the metropolis.

In the metropolis, mobility was far too great for it to be possible to establish majority rule based upon a unified group identity. Mobility and the multitude of identities forced a flexible political structure, characterised by temporary alliances between different interest groups. Politics was more about finding functional compromises than achieving consensus around an ideology. Certain trade-orientated and pluralistic nation states that developed early on, for example Switzerland and the Netherlands, developed relatively passive and pragmatic political institutions of a metropolitan rather than centralistic character. It is worth noting that neither of these states, nor any of the obvious metropolises, was affected by the revolutionary mass-movements which periodically brought about violent social convulsions in the centralistic nation states. This supports our thesis that the idea of political revolution is an integrated part of the overriding capitalist ideology, a symbiotic parasite on etatism and its nationalism, rather than an anomaly or an ominous portent.

Thanks to the fact that the capital took as its task the creation and management of the nation state with its bureaucratic and military apparatus, the metropolis was able to concentrate on trade, shipping and the colonisation of foreign territory. This co-operation and division of labour benefited both parties. The metropolises were responsible for international contacts, they received, developed and passed on impulses from outside, and, thanks to their banking systems, could be sure of stimulating the movement of capital that was necessary for the expansion of capitalism. The capital sent on extra labour and provided basic necessities, and the metropolis in return could offer luxury goods such as fabrics and spices. Besides this, the metropolis established colonial territories in distant continents in the name of the nation state. The monopolistic trading companies which several European countries established in eastern Asia are a typical example of this. In exchange for the metropolitan trading companies raising the nation state's flag on foreign land – which often happened in connection with the creation of brand new metropolises: Hong Kong, Macau, Singapore and Goa – they were permitted to exploit unchecked the new-found wealth of raw materials and virgin markets.

Because this co-operation benefited both parties, collaboration as a non-zero-sum game, a form of symbiosis soon developed between the capital and the metropolis. Intense traffic moved between the two types of city, a rapid exchange not only of goods and services but also of people and ideas. The capital stood for the exercise of political and military power: registration, cataloguing, legislation, as well as the production of ideology and collective identity; the metropolis contributed entrepreneurial skills, finance, trade, art, culture and individualism. It is important to bear in mind that right up to the end of the 1800s, it was the countryside which answered for almost all of the growth in population. Within the towns there were viruses and bacteria which weeded out any excess population. The epidemics which periodically swept across the world affected towns, mainly because of their population density and lack of hygiene, far worse than the countryside. This fact means that the countryside not only had to supply the towns with raw materials, but above all that it contributed a continual new supply of much-needed workforce.

The capital had a cohesive, stabilising function, whereas the metropolis stood for openness, new thinking and an experimental frame of mind which at times could be downright reckless. Apart from the concrete economic benefit these cities gained from each other, there was always a more subtle and mutual exploitation: they contributed to each other's identity and cohesion by acting as mutually demonised threats. The nationalistic ideologues and moralists could use the metropolis as the very embodiment of immorality and corruption. In a corresponding manner the leading tradesmen of the metropolis could ensure that the population remained in its place within the hierarchy, working hard to maintain the city's economic and political independence, by presenting the culture of the capital as intolerant and repressive. There was no real reason for these two types of city to work seriously against each other. The conflict that was encouraged was more of a symbolic and theatrical character, and actually had as its main purpose the reinforcement of those in power on both sides.

We have described types of city as purely theoretical opposites. It ought therefore to be pointed out that the great global cities which developed during the early stages of capitalism naturally showed characteristics of both capital and metropolis. But even where both variants collaborated within one and the same finite area, where settlements flowed together, if we look at the internal flow within the city there is a difference between capital and metropolis. The central square, with its parliament or council building, and the factory district on the outskirts, represent the capital. The harbour district, with its more or less sinful entertainment and artistic districts, thus represents the metropolis. In these melting-pot cities the boundaries between the two structures were both physical and mental, and often difficult to breach. Examples of this sort of hybrid city are London, St Petersburg. New York and Buenos Aires. Other cities belong, from an historical point of view, more clearly to one category or the other. Classical capitals include Paris, Madrid, Berlin, Moscow and Beijing. The metropolises include Venice, Amsterdam, San Francisco, Shanghai and Hong Kong.

If all of this development is linked to how the social interface of the average citizen during his lifetime grew from the church in the peasant village to comprise the flood of people in city squares, we can draw certain conclusions about how power structures and control mechanisms will be influenced in the transition from capitalism to informationalism, since the social interface of the individual citizen is now expanding enormously to correspond to the traffic on the global electronic net.

One example of how the "natural" connection between town and country in the nation state is disintegrating with the breakthrough of informationalism is the increasingly irreconcilable demand of the city that the countryside stand on its own two feet after centuries of subsidy. The nation, which was previously an invaluable asset to the capital, is now merely a handicap, an unjustified expense. In informational society seas and mountains are no longer required as "natural" boundaries and defences against the outside world, but as locations for recreation. The value of the raw materials that the countryside can offer in the form of wood and ore has collapsed, because the proportion of any product constituting the cost of raw materials has sunk dramatically while the value of the immaterial: ideas and design, have rocketed.

Population growth now takes place within the big cities while the countryside, in spite of ambitious regional political programmes, has been steadily and relentlessly depopulated. The war for market share between the different cities is no longer fought with weapons on land or at sea, but with information management as a weapon in virtual space. Under these new circumstances, a dislocation is taking place in the relationship between the capital and the metropolis: the capital is falling behind, while the metropolis, with its well-nourished banks of knowledge, its flexibility and its superior ability to build up boundary-breaking networks, is emerging as the urban ideal of the new age. The capital city's envy of the metropolis's freedom, mobility and independence is hastening the collapse of the nation state.

It is no coincidence that the most entrenched defenders of the nation state are the rural population which has most to lose when the strong nation is weakened and can no longer uphold its lofty promises. In the absence of any realistic negotiating position, pleas are being made for sympathy from the rest of the world, and for compensation for centuries of exploitation. Consequently the necessary but painful reduction of the gigantic subsidies which most industrialised countries pay to the countryside for the maintenance of their livelihoods is the great stumbling block during international negotiations on increased free trade. But the capital has little choice; the merciless competition between all the world's expansive metropolises which is the consequence of nodalisation has made it necessary for the capital, in a desperate attempt to transform itself into a metropolis, to dispose of its ever more burdensome responsibility for an entire nation.

The result is the denationalisation of informationalism, an increasingly rapid deconstruction of the nation state, a sort of global perestroika for nationalism. This development is reinforced by the fact that decisions within the capitalist system are made through elections, and that a majority of voters lives in cities, while, at the same time, informational society is controlled by networking, something the metropolitan part of the urban population has conscientiously practised for centuries. Squeezed between the power structures of both paradigms, the countryside does not stand a chance. Not even democracy can save rural subsidies in the long term, because the rural electorate is a minority, a group of opinion which is steadily shrinking and becoming easier to ignore. With a radically dwindling income from taxes for the state as a result of the expansion of the virtual economy, the urban population will use its majority position to vote down any further support for the countryside, which is therefore being decolonised and left to its fate.

Even the military is retreating from the countryside during the last phase of capitalism, which can only be explained by the fact that there is no longer any territory left to defend, still less any desire to defend anything. Territory is reduced to an area for environmental politics. Carbon dioxide levels are measured, the acid content of bathing lakes is neutralised, and there are discussions about whether the countryside should be open (as under capitalism) or covered with forest (as under feudalism), and about who should pay for it all. The question of which flag happens to fly in the wind in some distant end of the country is not something which keeps urban inhabitants awake at night. During informationalism the leading capital cities will no longer compete for how much territory they control. On the contrary, they would sooner be rid of the responsibility for as much territory as they can, in order to make the grade and join the leading metropolises of the world. This development is most noticeable in Europe and east Asia, because there it has been reinforced by a declining birth-rate.

In the aftermath of these collective convulsions, it is inescapable that the self-image of every individual is going to undergo a radical change. The breakthrough of informationalism means that the Enlightenment idea of the perfect Man as a replacement for God, and therefore the idea of the need to realise one's "true self", will finally be laid to rest. Instead, there is a vision of the body's liberation from predetermination, the possibility of switching between different identities, rather like changing clothes according to situation and context. The subject will find him or herself in a state of permanent becoming, always receptive to new impulses, a continual system of evaluation without a clearly defined goal. The ambition will not be to achieve something finished, thanks to effort and discipline, but rather to keep alive as many opportunities as possible. We are talking here, to borrow a phrase from Deleuze, about a dividual rather than an individual. This dividual does not have one identity, but many, and is constantly divisible. This is consistent with the biological fact that it is impossible to localise the ultimate instance of control within the self. Man is recognised as a meeting place for a mass of contradictory desires and powers, with no centre.

Technological development is forcing a new mobilistic identity onto man, by pulling the carpet from beneath the totalistic ideal. On the net the individual's identity will arise in the context that is actual at the time, only to perform a breath-taking transformation the next moment. Out goes the intact individual, chained to his identity as to a heavy rucksack, and in comes the free- flowing dividual. In a first phase the dividual will stop trying to be Man, then in the second phase it will become impossible for him to return to being Man, however much he may want to. This new freedom both attracts and frightens, but it cannot be avoided. What we are describing is the virtual subject of the plurarchical, and hence post-democratic, society; where everyone decides over him or herself, where no-one can decide over anyone else in the name of a superior majority.

It would be a big mistake to confuse this plurarchy with anything resembling anarchy. Plurarchy is not the same as the abolition of rules, where everyone can do as he or she pleases. In plurarchical society there are actually more rather than fewer laws and rules than in a democracy, more complicated, abstract, and, to a greater degree, elevated out of sight and beyond discussion. The plurarchical rules of the game are, however, not determined by political or judicial institutions of the nation-state variety: the courts and parliament will quickly see their status and power dramatically devalued. Instead, the rules of the game on the net will be meme-Darwinian, the intricate system known as netiquette, and it will be this that characterises the strict ethics of informationalism, and which will increasingly replace the laws and regulations of the capitalist paradigm.

These rules cannot be permanent in their nature. Like everything else in informational society, they must be constantly updated because their complicated context is in permanent flux. Netiquette ought therefore to be understood as a living document without given limits, a sort of quasi-judicial organism in perpetual motion, a net within the net which both forms and at the same time reflects cultural values and ideology. Crimes against netiquette will not be punished like crimes against laws in the old paradigm, with prison sentences and forced internment in other institutions, nor with fines or economic punishments: treatment like this is not sufficiently recognisable as punishment in a society where life is mainly lived on the net.

Anyone transgressing netiquette will instead have to look forward to virtual displacement: expulsion from essential networks. Like anyone breaking the norms of the nomads of the savannah or the villages of feudalism, anyone breaking the norms of the electronic tribe will be punished with exclusion from the community of the group, which will lead to a dramatic loss of social identity. In milder cases it might be sufficient with network harassment for a determined length of time, but more serious or repeated crimes against netiquette will lead to virtual isolation. A parallel to this system of punishment is the trauma caused by enforced unemployment in capitalist society. Add to this an inability for the unemployed to find any imaginable alternative occupation, and the extent of the effects of network exclusion becomes clear.

It is the curator, the overseer of the network, who will shoulder the role of legal administrator in informational society, and who imposes and enacts punishment. Within the curators' own network, informational society's equivalent of Interpol, there will be a constant flow of information about which net-citizens are undesirable within attractive networks for the time being. Only in so far as it is possible to maintain vital competition between different curators and their different networks can the plurarchy function; otherwise the whole of virtual society is threatened by an oligopoly of the curatoriat, a virtual form of minority rule with a strong concentration of power and with great scope for corruption and arbitrary behaviour.

But at the same time, the citizens of the net will keep control over themselves to a remarkably large extent, because flexibility and adaptability will pay in the informational system of punishment and reward. If you study the historical development of modern society, one of the very earliest noticeable trends, as identified by the German-Jewish sociologist Norbert Elias and others, is a constantly increasing degree of internalisation. This means a process by which the prohibitions which previously had to be expressed, written in law and maintained with the help of severe punishment, come to be regarded more and more as self-evident: repression has moved from external social institutions to within the consciousness of the citizen. Novelists like William Burroughs and philosophers like Foucault and Deleuze have been interested in how the maintenance of social morality during late-capitalism has moved from being a question of state-organised discipline to one of internalised control. One consequence of this development has been that commands and directly expressed threats have been increasingly replaced by pedagogically expressed propaganda, whose purpose has been precisely to educate the citizen to put in overtime as his own moral policeman. The control society has replaced the disciplinary society, treatment has replaced punishment, and observation has been delegated to the person under observation. It is much cheaper that way.

One condition for this control society has been that the myth of Man has been kept alive, that each individual citizen has recognised this ideal and has constantly been reminded of the fact that he has not been able to live up to the demands that have been made, and that he therefore has every reason to continually improve himself in every possible respect. To become Man has been a lifelong project which no normal person has ever managed to accomplish. The myth of Man has proved to be extremely tenuous of life, which explains why the elite corps of capitalism has, despite everything, succeeded in keeping power over a society which has rapidly developed towards a plurarchy, and where the ideological legitimacy of power has largely been eroded. Kidnap victims often display irrational connections to their guards – the so-called Stockholm Syndrome – and the same symptoms have been observed among the consumers of late capitalism: a mysterious inability to let go of the dated and deforming myth of self-realisation and "the true me".

During the final period of capitalism, state organisations and various research institutes have laid out a formidable minefield of alarming reports where the increasingly deficient morals of the citizens are illustrated with tables and diagrams. Everything from obesity and lazy dietary habits to tv-viewing and a lack of empathy have been depicted in strident tones, which have then echoed even louder through the megaphones of the mass media. The immorality that is perceived as being spread by the Internet has been particularly problematic, for obvious reasons. Each problem has been worse than the other, and they all require powerful initiatives in the form of information campaigns from state organisations and various research institutes, which are consequently in great need of the allocation of massive resources. Little man, as all studies obviously show, has not proved mature enough to fulfil the sacred task of realising his true self. He therefore needs masses of tender loving care and direction. The result has been a bizarre campaign of slander where the dying capitalist institutions: the political classes, psychologists, socionomists, the mass media, school and family, are all accusing each other of a lack of responsibility for the education of the people: Why will no-one take responsibility and try to stop people taking responsibility for themselves?

This intimate collaboration between politics, research and the media has been so intricately wrought that every hint of a criticism has been organised out of existence. It has certainly not been in the interests of the headline-hungry media to subject any of these alarming reports to close scrutiny. And the commercial powers have been able to watch the entire spectacle as a single huge advertising campaign for all manner of ego-boosting products and services. A citizen who is dissatisfied with himself is generally a high-achieving consumer. Perfection is an impossibility, but there will always be new products and services to try out, proclaim the mass media through their one-way communication, and to try to improve oneself at the very least is one's duty as a citizen. No-one is thin or beautiful or well-dressed enough to be left in peace from the constantly repeated incitements to improvement. The perpetually piled-on self-loathing is the very basis for the therapeutic hyper-consumption which is keeping the late-capitalist wheels turning at such a rate. Advertising is the carrot and the shock reports the stick. State and capital are sitting lashed together in the same boat.

In informational networks, where communication is interactive, capitalism's old truths about the individual will be subjected to hard critical evaluation. The myth of self-realisation will be seen through, and the whole idea of a unified self to nurture will appear antiquated. As a result the political, academic and commercial forces behind this propaganda production will gradually lose their influence over the collective consciousness. But this certainly does not mean that the terror of overheated commands is a closed chapter, it just means that new "truths" will be packaged differently. The netocracy and its higher functionaries, the curators, step into the arena and lay out the new, constantly changing text of the law, constantly new little netiquettes for the new dividual. Even if plurarchical structure makes it possible for every single player to question the single curator's current set of rules, there is no alternative other than to submit to the rules of another curator, and the differences between them might well be negligible because the curators in turn have organised themselves into a meta-network to defend their power as a group.

The rules will change, but the constant underlying message of the curators to their net-citizens will be simple and unambiguous: you can never network well enough, you can never be good enough at communicating, you can never let yourself rest, you must constantly be ready to jump, constantly ready to learn new things. Thus a new set of masters will seize power and the language of power with which to control informational society. The definitive difference between a capitalist strategy and an informationalist strategy for power is that the curators understand the consequences of interactivity and how this is indissolubly linked to power. So the curator will not speak to his subordinates, he will converse with them. The netocrats in power are listening rather than commanding, they communicate discretely and subtly instead of brutally giving orders. Netocratic power is not the power to take decisions as such, because in plurarchic society decisions will be taken by the individual concerned, but control over the understanding of the consequences of alternative decisions.

In informational society the function of the individual net-citizen as his own moral policeman is no longer prompted by an arrogant and effectivity-maximising nation state's guardian-bureaucrats, but instead inspired by constantly changing netocratic trends which determine what it is that makes someone "hip", and how to show that you've got "it". Ethics will become more and more a question of aesthetics. "Poor style" will become synonymous with social suicide. Every single player in a network will always be assumed to be fully aware of netiquette, the unwritten laws for how members ought to behave towards one another. Because net-competency is the key to success in the informational society, there will be a rich blossoming of all sorts of communication therapies and courses to teach hopeful networkers what is "good netiquette".

The problem is just that because we are talking of dynamics, new knowledge will soon be old. The truly terrible thing about these rules is actually that the more widespread they become, and the more people learn to live according to them, then the sooner they will lose their value. The most acute threat of inflation in informational society will not affect currencies or securities, but the value of each single player's actions. The single net-citizen's social interface is global, his access to information practically unlimited. His social identity is an open question, and his position of power is ultimately a question of style.

9. Network pyramids – attentionalistic power hierarchies

One of the fundamental concepts in the discussion of informational networks is transparency: the network as a translucent, pellucid, and therefore equal and democratic system. The principle of transparency is that all the members of the network have access to all relevant information, and that at any time they wish they can make their own contribution to the internal debate. All cards are on the table, everyone can form their own opinion and make comments, everyone can participate in the decision-making process. If transparency is the dominant principle in informational society, this will entail a revolution in the labour market. When a traditional business is transformed into a transparent network, the employer can no longer treat his employees according to feudal principles governing the exercise of power. Everyone's eyes are on him. A Machiavellian cannot act in a Machiavellian way when he is constantly under the gaze of his subordinates from every direction.

This is the background to late-capitalism's sudden blaze of interest in ethics, which should be regarded as a textbook example of gallows conversion. What is to be done? When despotism and capricious behaviour no longer work, new strategies must be found for leadership and the exercise of power. Technological development is driving society and is giving, in this particular instance, the individual employee previously unimagined influence over his work situation. The networks' internal transparency and the workplace democratisation connected to it has led management theorists to proclaim triumphantly that "Marx was right": development of the conditions for production has made old production practices obsolete and untenable, the workers are assuming power. But there is one snag in this rose-tinted scenario. The problem is that informational society is considerably more complicated than that, businesses are becoming virtual networks to an increasing extent, with fewer stable structures and rigid employment practices. The dynamic of the network is a phenomenon with considerably more aspects than mere transparency, and of greater importance as well.

For the network to live up to the harsh demands of effectivity that exist in the curators' market, everyone who is not regarded as contributing anything valuable, or who is simply perceived as a threat to the network members' common interests, must be unconditionally placed or kept on the outside. Every network with the slightest ambition to be attractive and successful must thoroughly sift prospective members, otherwise it will soon collapse under the torrent of irrelevant information that will flood in and fill up the limited amount of space. Development like that will inevitably lead to the key members losing interest and allowing themselves to be enticed to join other networks where the policy is more restrictive, whereupon the network they leave behind them will be transformed into an empty shell doomed to become extinction, full of impotent actors who, in confusion, are producing irrelevant nonsense.

The open networks that have been made possible by the rapid expansion of the Internet will either be transformed into closed networks, or will fall into disrepair and become gathering points for worthless garbage-information. In closed networks, the members are hand-picked by the human curators, the doormen of the virtual world. Tall, thick firewalls will be constructed around the networks, to protect against unauthorised observation and undesired entry. The more attractive a network is, the more people will want to become members, and the higher up the power hierarchy the network wants to get, the tougher the entry requirements will be, and the more insurmountable its firewall will need to be.

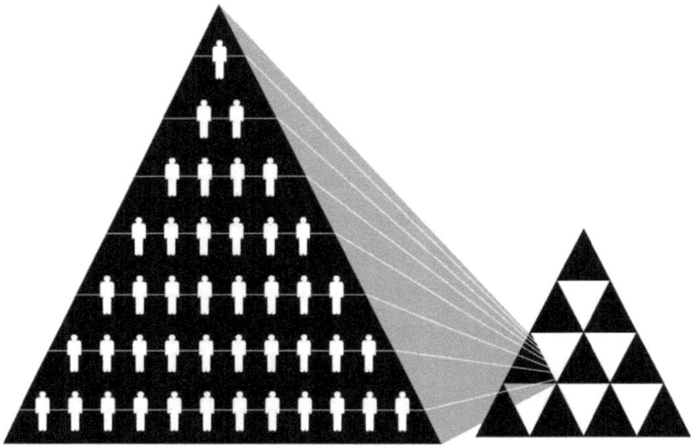

fig.9.1 Network pyramid

The direct consequence of this dynamic is that virtual society is forming itself into a long series of network pyramids, a power hierarchy where consumtarians are directed towards the least attractive networks full of garbage-information, while the netocrats lay claim to and themselves define the highest networks where power and status are concentrated. This society is by definition post-capitalist, because the requirements needed to achieve status under capitalism – money, fame, titles, and so on – no longer have any value for entrance applications to any of the higher and more powerful networks. The netocratic status which is now in demand requires entirely different characteristics: knowledge, contacts, overview, vision. In other words: qualities which contribute to increasing the network's status and making it even more powerful.

Capitalism's perception of power-structures, that all valuable human activity stems from and is controlled from a central core, has also become obsolete. There is, quite simply, no centre in the virtual world. The netocratic network pyramid is not primarily constructed with the aim of exercising power, it is actually not constructed at all in any usual sense, but should be seen as the structure that technology itself has promoted according to Darwinian principles of natural selection, and which renders impossible the control necessary for protection of certain specific interests. The principle of the network pyramid is more decentralisation than any centralised concentration of power. It will never achieve equilibrium, its power relationships are constantly changing, which means that the power that is exercised stems from temporary, nebulous, unstable, mobile alliances rather than from any particular geographic point or any particular constitutional entity. Power will thus become incredibly difficult to localise at all, and therefore naturally even more difficult to criticise or combat. But the fact that power becomes more abstract and invisible does not mean that it vanishes or is even weakened, but rather the exact opposite.

The social status of the networks is determined by how well their professional doormen, the human curators, who are just as much arbiters of taste and supervisors of netiquette, perform their difficult task. Because the curators in turn are competing within the market of the networks, they will constantly need to advertise their own network and convince other people of its excellence, partly to entice attractive new members, and partly to be able to enter into influential alliances. Each successful decision one way or another might lead to spectacular gains, while each unsuccessful decision about access or exclusion might have devastating consequences for the curator's, and therefore the network's, reputation and credibility. It is reputation, or capital of trust, which is the network's most important asset; with the help of reputation, attention is attracted to the networks, and there is a great shortage of attention, rather than money, on the net. Money will follow attention, and not vice versa. Attention is the only hard currency in the virtual world. The strategy and logic of the netocracy are therefore attentionalist rather than capitalist.

The individual curator's existence is turbulent and uncertain, but the group of curators need hardly worry. Their collective position of power is unthreatened, a circumstance not unconnected to the fact that the curators' activity cannot, for the foreseeable future, be carried out by machines, and is not regulated by any standardising rules. There is no accumulated wisdom, no ideological tradition, no precedent to call upon, no education or exams to fall back on. There are, quite simply, no formal criteria to refer to. This means that none of the old public institutions from the old paradigm has in the long term any prospect at all of successfully competing in a developed network market, where fingertip sensitivity, intuition and stylistic consciousness, excellent social skills, will be more in demand than anything else.

One consequence of this development is that the most sought-after and therefore most valuable information of all on the net is that which concerns networking itself: how to construct and administer your network in the most intelligent way. This means in turn that the most powerful network of all is the meta-network where the curators cement contacts with, learn from and enter into alliances with each other. The global meta-curatoriat, the highest network in the curators' own network pyramid, is the most powerful institution of netocratic society, informational society's equivalent to a global government. But the fact is that the whole system is fluid, which means that its constitution will constantly shift, making an empirical analysis of power more difficult.

Informational society presents an entirely new topography. Under capitalism, despite the slower and generally more limited communications, there was still a considerable degree of clarity, as a result of the fact that circumstances and the overriding logic were largely the same throughout the system. In this way capitalist society can be said to resemble an open field where people can see each other, admittedly at a distance, and communicate with each other, even if it is not always easy to hear. We call this common area of interests "the public arena". The topography of informationalism, paradoxically, is more like a labyrinth. The arena is crooked, and events are unpredictable. Round every corner there is a complete surprise. What was relevant yesterday will seldom have any relevance today. It might well be necessary to abandon a successful strategy without any warning at all. The transparency is essentially a chimera, a netocratic propaganda myth, existing only within extremely narrow, horizontal sections. We can see and hear everything more clearly over short distances, but the cost of this is that our far-sight, and with it the public arena itself, has disappeared.

When an excess of information is produced, attention is at a premium. In the netocratic network information itself is of limited value. In contrast, there is a value in being able to avoid unnecessary information, in order to free up valuable time and facilitate concentration. The information which is sought-after must be relevant and reliable, and preferably also exclusive, and this particularly valuable information can only be found within the highest networks. And it is only in the highest networks that there is the knowledge and overview necessary to use information in an optimal way. Meta-information, about how different types of information can be connected in the most effective way, is in itself the most valuable form of information.

The sale of first-hand information to the highest-bidder at some sort of public auction will in time become an increasingly unlikely scenario, for several reasons. Exclusivity declines if the business transaction is public, and the risk for costly leakage from the highest-bidding network so great, since attention and not money is the strongest currency, that information will eventually find its way to the highest ranking network in any case. For the netocrats, who you communicate with is simply more important than how much you get paid for the information you have to offer. In the higher networks a capitalist strategy will only work in the extreme short-term, because the social costs of further sale of valuable information will, in the long term, be unsustainably high.

Survival among the netocrats demands that you acquire long-term, attentionalist thinking based precisely upon the insight that attention has a higher value than money.

Capitalists will become an underclass which has to content itself with haggling over old, second-hand information from the scrap-heap, while the netocracy – the networking elite – carries off the prize of power and status, as well as experiences and kicks. The netocrats will, of course, also eventually carry off the financial profits, in spite of the fact that these are of only secondary interest; they control the knowledge, after all, and can create the attention that is more valuable than anything else. Only when the netocrat has used the most important information for his own advantage will he sell it, with the blessing of his network, to the highest-bidding capitalist. So the capitalists will be forced to adapt and subordinate themselves to attentionalist conditions and play a game whose rules they have not decided themselves, and which they in many cases do not understand at all. Already during latter-day capitalism this pattern was becoming apparent when the capitalist with the best network behind him beat the capitalist with the most capital in the struggle over who was going be allowed to invest in the new economy. As a result, the capitalist economy had already been subjected to attentionalist principles.

During the initial period it is possible that certain human curators will try to sell their power for money. But this short-term action will soon turn out to be self-destructive, because this sort of transaction will rapidly undermine the curators' own credibility, which means that they will soon have nothing left to sell when the standard and status of their networks has plummeted. As a result they will inevitably lose the power they once had, and be knocked out of the extreme competition that exists within the curators' market. It is clear that the netocratic power that can be bought for money is not worth it, since it will run through your fingers like sand the moment anyone tries to capitalise on their purchase. Thus money has an indisputably inferior value in the power-hierarchies of informational society.

The dynamic that already exists both within and between the different networks is forcing an entirely new prioritisation in the game for status and power: careful care of one's own trademark. The players simply cannot afford to be seen in any sleazy old context, because then they would lose all credibility. At the same time, it is invaluable to be seen in the right context, to belong to the truly interesting alliances, something which can no longer be bought with money, because money is no longer interesting enough. Therefore it is no exaggeration to say that the paradigm shift is bringing about the death of capitalism, simply because capital is being forced to court attention, instead of vice versa.

Attention, which there is such a shortage of, is what every player wants. According to fundamental national economic principles of supply and demand, it is no longer financial profit which drives development and motivates people. Old values are being badly devalued. This is the case with both money and titles; it does not matter whether these titles are connected to family, politics, business, or education: their value is sinking. Attention is superior to everything else; society is attentionalist rather than capitalist. If the word informationalism were not already established within sociology as a collective term for the new paradigm, the word attentionalism would possibly be even better for capturing its most characteristic feature. It is thus of crucial importance to be clear about what attentionalism means, and what the unavoidable consequences of its inherent dynamics are.

The Internet differs from latter-day capitalism's mass media on one conclusive point. The daily press, radio and television are all examples of one-way communication. The consumer consumed, while the transmitter sent out his message in peace and quiet. A closed system without dialogue, without a forum for criticism or questioning, apart from the strictly managed contents of the letters page. The net, on the other hand, is about communication in at least two, and often even more directions. Before, the medium was the message, but now it is the user. By acting on the net, the user is creating content: the boundary between production and consumption is dissolving. This means that those in power under capitalism, the possessors of the much-vaunted hegemony of public space – politicians, propagandists, preachers – are no longer serious players in the media arena. They are now being deposed by finely-tuned network consumers, who are constructing increasingly sophisticated feedback loops of information. These are then thrown back into the system and undergo a long series of manipulations in the hands of the many participants. It is a matter of spectacularly complicated processes with infinite variables and often astonishing results.

The fact that the new network citizen is not particularly interested in old, worn-out rituals from the capitalist paradigm, like for instance political elections, is therefore not particularly surprising. The audience has simply got up out of its chairs and left the theatre. The old capitalist is left alone on the stage asking himself more and more indignantly why no-one is listening to his words of wisdom. From the foyer there is an increasing noise: the audience has started to communicate amongst itself. Someone is offering drinks, and from the loud-speakers in the corners dance music is playing. In this half-lit night-club world it is the host of this improvised party, the curator, the virtuoso networker, who is the big new star. Power in informational society does not belong to the person who dictates, nor to the person who believes that the centre is in the spotlight, but to the person who sets the social game going with discrete effectiveness. What is decisive is not what is communicated, but how it is communicated, and who is communicating with whom. The aim is not so much an aim as a direction; to create, maintain and strengthen a process in motion. The ambition is not to shout louder than the media hubbub, but to use the wind-machine as a musical instrument on which the netocrat can play a seductive melody.

The displaced balance of power that is the result of this development is reflected in the growing medialisation of politics. Politics is being forced to adapt to media dramaturgy and become tv-friendly and entertaining in order to attract any attention at all. In the long run this will lead to politicians losing all real power; they will become a new category of low-paid tv-entertainers, harmless enactors of other people's scripts, mercilessly reviewed the whole time, and mocked by their journalistic overseers. There is no alternative. Anyone trying to resist developments and persist with political agitation in the old way would not survive in the society ruled by the netocracy. He or she would look like a despicable information-tyrant. The new conditions of the informational media landscape mean that the plurarchic public is turning its back on the old political stage.

The players' own ambitions, what or who he or she has decided to be, are no longer of consequence. No-one "makes their own fortune" any more. The whole of the perceptual world reflected in this phrase has passed away. What matters instead is the player's capacity for instant conversion the moment the requirements and desires of his or her surroundings shift. The incredibly complex and constantly changing system of rules for virtual behaviour which is now being developed and endlessly updated is shaping and expressing the netocratic ideal. The only useful key to impenetrable netiquette is maximal flexibility and highly developed social intelligence. Nothing can ever be taken for granted; all our prejudices, all the fictions that we use to orientate ourselves in life must be constantly re-evaluated. Every new situation requires a new judgement, as well as the courage to act without hesitation according to this judgement, and to behave accordingly. Only those who are prepared to allow themselves to constantly change can survive socially in a world which has realised its own constant mutability.

This has a number of interesting consequences. The netocrats' therapy sessions, to take one example, will no longer be based upon prevailing psychological or psychoanalytical models of a more or less integrated, unified ego. Freud and all his followers, interpreters, successors and adversaries within the therapeutic genre have played out their roles. Instead the situation- and interaction-based ideas about social intelligence and controlled exploitation of personality-division of progressive communication theorists and schizo-analysts will come to dominate the field. In informational society more people will seek treatment for the many problems connected to an excessively unified personality than for any difficulties connected with schizophrenic tendencies. It is almost the case that a manageable form of schizophrenia is a netocratic ideal.

The netocrats are not interested in realising themselves or establishing contact with their true selves. This sort of concept is, in their eyes, old nonsense, superstition. They neither believe nor want to believe in what they perceive as the social constructions of a bygone age. Instead they strive to nurture and refine their capacity for simultaneity and the art of constantly developing a multiplicity of parallel identities. Personal development is the realisation of all one's possibilities as a dividual, a pragmatic alliance of essentially different temperaments and personality traits. The old individual will look like a feeble, one-dimensional wretch rather than the ideal. Schizophrenic, kaleidoscopic identity is in contrast exemplary in the sense that it is functional. Schizo-analysis will contribute to strengthening the capacity for constant change according to constantly changing circumstances.

Communication in informational society can only take place within systems that allow feedback. Old, one-way communications will become curiosities in museums. The very concept of communication presupposes not only a recipient but even a more or less qualified reaction which returns to the original transmitter and functions as incoming data in the revolving process of the feedback loop. It is in this rapid interplay of interactive information management that the many facets of the dividual glint and glimmer. All the various dividuals illuminate, determine and confirm each other, in restless motion yet extremely co-ordinated, like fish in a shoal. The affirmation of the dividual rests in the reaction of its counterpart, or counterparts. He or she will regard him or herself as a network dependent upon other networks in order to use a constant flow of information to continually update his or her keenly desirable virtual identity.

The netocratic dividual will seek rest on a platform, only to discover at once that the platform in question is in motion. He will then step onto another platform on another level, once again in search of firm ground under his feet, only to discover once again that that platform is in motion. So the process goes on, motion in different directions on different levels. But the receptive dividual will soon turn stepping from one platform to another into an art-form, and continue his search for firm ground despite knowing that the project is illusory, yet fascinated by his virtuoso abilities. The floating state which arises in an environment of consciously constructed and constantly revised fictions will replace belief in firm ground. Consciousness of the fragility of fictions will lead to a disillusioned clarity and loss of meaning, but also to a creative intoxication of freedom and limitless possibilities. When fictions are created in interaction with other dividuals the possibility for communication arises, and then anything can happen. This is the floating point of departure for mobilistic thought which is the only perceptible path through informational society.

One distinguishing characteristic of informationalism is the escalating medialisation of society, which is propelling all previous reasoning about credibility to its limit. A pedagogical way of explaining this process is to use a classic mass-medium, for instance a weekly or monthly magazine, to illustrate what this general medialisation entails. When we buy a monthly magazine, we do so to get a certain amount of relevant, substantial, engaging or at least entertaining information, what the netocracy calls content. But the magazine also contains a large amount of information that we do not want, but which we have to accept, at least fleetingly, in exchange for getting the desired information at a lower and manageable price, in other words: the magazine contains adverts. And the adverts bother us. A monthly magazine with too many adverts in relation to content cannot hope to retain very many readers. Besides this, a magazine that does not differentiate properly between adverts and content loses credibility and therefore readers. A magazine with no adverts at all is the one with the highest credibility, if not necessarily the most readers. At least if this is the result of a conscious policy and not merely a failed attempt to attract advertisers.

In a completely medialised society it will become necessary to apply this thinking not only to media consumption in its classic form, but also to all social activities in general. In the virtual community on the net the difference between content and adverts, in the sense of purchased information, must be constantly maintained; the battle for attention never lets up, but cannot merely be described with capitalist logic. Competition between the new people in power, the distributors of information, the curators, is so cut-throat that any curator who has the wherewithal and the integrity to limit the purchased information to an absolute minimum will be the one who accumulates greatest credibility for his network and therefore the one who emerges victorious. Purchased information, the only information that capitalism is interested in or even has the capacity to understand, has a mainly negative value to the netocracy. It is only asked for by the consumtariat who are looking around in confusion for instructions for consumer behaviour. In this way the capitalist renders himself impossible in netocratic circles, his activities disqualify him from membership.

In the attentionalist culture which is taking over from capitalism it is thus access to relevant and exclusive information, much more that access to money, which is decisive in the distribution of status and power. Innumerable investors will gather around every informationally wealthy person or organisation, hunting for information, begging and pleading to be allowed to invest their financial capital in the activity which he does not himself understand, and they will be forced to adapt to the various wishes and whims of the netocrat. The Fortune 500 will be reduced to a sort of kitsch entertainment literature, rather like Debrett's under capitalism; as a piquant diversion, one can study how a bygone era ranked its citizens according to long-lost principles. Other measurements apply under informationalism. It is now network-membership that determines a person's social status, and this is constantly changing, both with time and depending on context. A list of the aristocracy of the new paradigm could never be a printed book, because it would be immediately outdated, and because anyone would be able to read it. Netocratic relationships are mapped out online instead, constantly updated and hidden behind codes and virtual keys, for use by selected party-organisers and other curators.

Invaluable contacts are literally invaluable: they cannot be purchased for money. Contacts only become accessible through the transfer of other contacts of equivalent total value. Attentionalism will therefore create a sort of sophisticated bartering system. Contacts will be established when a player has valuable information to pass on, and, above all, when the person concerned exhibits an exceptional capacity to manage and present the information in question in an appealing way that arouses interest. Here yet another distinguishing feature of attentionalism becomes apparent: it is possible both to have your cake and eat it, you keep the information yourself, and its value, at the same time as sharing it with a select few. The fact is that value is accrued in connection with the transaction: communication creates attentional value for the person offering the information, and that value remains so long as the information in question is of interest. The value will also persist for as long as there is any reason to expect more of the same product from the information source.

The information you choose to keep to yourself obviously has no attentional value at all, and the information you sell for money has an extremely short-lived and limited value, whereas valuable information that is beneficial to the network increases the attentional status of the informer enormously. This is the principle of the internal life of the network, a micro-medialisation on the level of the transparent network. It is within this process that the individual player can climb socially, it is within this process that the network ensures a regular supply of valuable information. It is from the players who play this game to its conclusion that the highest networks recruit the occupants of their formal positions. The consumtariat, on the other hand, is directed to scratch its living on the virtual market of temporary occupation; they are project-nomads who are constantly hunting between the employment-offering homepages.

The faith in progress that was a characteristic of the capitalist paradigm and which was expressed once again in the clueless technological optimism of latter-day capitalism – everything will be alright, the Internet is a vitamin injection for democracy and everyone will share in the benefits of the new economy – appears to the netocracy to be a vulgar substitute for religion. This does not stop it from being an important element of netocratic propaganda; religious needs are constant and the consumtariat needs to be tranquillised with a substitute for traditional expressions of faith when secularisation increases. The consumtariat must therefore be kept in a good mood with constant new varieties of spectacular entertainment, which is synonymous with success on the consumtariat's level. But within the netocracy the idea of progress will be replaced with extreme mobilism and relativism.

Networks will blossom with lightning speed, capture power and status rapidly, only to disappear just as quickly. Nothing is permanent, no value lasts, all platforms are in motion in different directions. Admittedly, from the perspective of evolutionary sociology it is possible to discern a historical trend; the broad sweeps are all pointing unambiguously in the same direction: growing globalisation, ever more sophisticated non-zero-sum games, increasingly complicated networks of mutual dependency. But this is not the same as progress; new technological and social circumstances may well create new winners, but also new losers. Some social problems become less acute, while others more so and new ones appear.

Rapidly blossoming networks should equally be seen as a so-called "event", as an organisation with a enduring temporal presence. Intensive activity in temporarily constructed feedback loops will create social energies that are difficult to harness. The resonance is multiplied many times over through feedback; the pressure waves spread quickly through the whole body of society and therefore the meme-Darwinian preconditions for other networks and their players are altered. In the same way that an organism's surroundings in nature essentially consist of other organisms, so a network's surroundings are basically other networks. There is a perpetually ongoing race – one predator gets stronger, its prey have to counter with ingenuity or speed, and its competitors among the predators have to come up with a counter-reaction; network X goes on the offensive, whereupon networks Y and Z have to come up with counter-offensives. Every change has consequences that are infinitely difficult to foresee (the variables are infinite), but nevertheless fully logical. Nothing is more important to an understanding of informational society than knowledge of these mechanisms. Therefore analysis of network dynamics will be the social science which attracts the greatest interest in the informational paradigm.

In this way, the Network will replace Man as the great social project. The network of curators will replace the State as the visionaries, managers and ultimate power of the great project. Netiquette will replace law and order as fundamental human activities increasingly move into the virtual world, at the same time as the authority and power of the state diminish as its income through tax shrinks and national boundaries dissolve. The curator will assume the state's function as moral guardian and will punish all transgressions against netiquette by expelling the delinquent from attractive networks. Exclusion, restricted access to information and other forms of rescinded membership privileges are the netocracy's methods for deterring and controlling dissidents. Severe limitations of virtual mobility will be the informational equivalent to internment.

In order to protect their common interests and make the administration of the networks more effective, the curators will establish their powerful meta-networks for policy questions about net-policing. Note that the curator acts as policeman as well as prosecutor and judge, and that no formal legal rights can be created in the changeable system of networks. The deliberations that take place within the highest networks will be protected against every intimation of observation, and need not stem from any democratically reached decision or any generally embraced traditions. The entire principle of public life and the public's right to observe generally will become extremely difficult to maintain; in spite of all optimistic hopes to the contrary, the public arena will disappear and be replaced by the labyrinthine topography of informationalism. A general perspective of overview will be impossible.

But informational society is still not totalitarian. It will not, admittedly, be possible to appeal against the curator's decisions in the usual sense, and the chances of getting justice for wrongly meted-out punishments will not be great. But the extreme mobility and multiplicity within the system will ensure that there are always alternative networks to join, so long as you really do have something which is attractive to the market. If you have, the other networks will not be able to afford to decline and refuse membership, nor will they have any reason to. And this complex topography will affect every player. Even the most powerful of the curators will lack a complete overview of informational society and will therefore already be exposed to a limitation of his potential power.

What we know today as capitalism will not vanish, just as feudal structures were not destroyed with the advent of capitalism; they simply carried on in a weaker state, subordinate to the victorious paradigm and its logic. The same thing applies now: a capitalist pattern will form an important component of a superior, informational system. Capital will become, to a very great extent, something sought by the consumtariat, and money will be the language used to describe and measure traditional consumption of goods and services. But the fact that money has been digitalised, that the movements of financial capital are instantaneous and impossible for politicians and bureaucrats to regulate, will disadvantage to a greater extent than before an underclass without access to relevant information, at the same time as it benefits a well-informed netocracy with its finger on the pulse of the flow of capital.

The fact that speed is increasing and the markets' complexity is growing means that the information abyss between the initiated and the excluded is becoming wider and wider, at an increasing rate. When the information that was once attractive has filtered down to the lowest-ranking networks – after a few minutes or several weeks, in so far as it filters down at all – it will be so old that it is worthless. The possibility for action, for instance to make gains from moving money from European bonds to some obscure growth market in deepest Asia, has long since vanished. Information will only have been interesting, and only had any real value, so long as only a chosen few knew about it.

Of course, the netocracy has ensured that it has the services of leading experts, indentured underlings providing discrete and effective protection of the new power's interests. But in spite of this, for the netocracy power and status are still a matter of something completely different, namely exclusive information that cannot be bought with money, the sort of information that would be exchanged for blank cheques if you were in the mood, and if cheques were still around. What is important is the capacity to create attention within the circles that matter. Having something to say which gets the noise of information to fall silent. Welcome to attentionalism!

10.
Sex and tribalism, virtual education and the inequality of the brain

Trends have a distinct message for us. The fact that information and communication technology is breaking through a high threshold and entering a new historical phase means that everything is changing. We are being forced to see ourselves and our surroundings with new eyes. The paradigm shift will have wide-reaching and very real consequences when our old "truths" lose their validity. The institutions which supported the old society, and which seemed under the circumstances of the time to be eternal and "natural", are now suffering extreme crises and being revealed as products of a society and an ideology that were intrinsically tied to the circumstances of their period. Their rapidly approaching demise under the crushing weight of change therefore appears inevitable. This trend is inescapable. The institutions most affected by this are the nation state, parliamentary democracy, the nuclear family and the education system.

The concept of a "trend" ought not to be understood as synonymous with fashion or anything like that; it has nothing to do with the world of glossy magazines. We are using the words trend and countertrend in analogy with Friedrich Nietzsche's conceptual pairing of action and reaction. For Nietzsche, action was connected to the desire for power; it is, in a philosophical sense, an original impulse, independent of other impulses. Reaction, on the other hand, is a secondary impulse, and only arises as a response to action. It is quite literally reactionary in the sense that its main aim is to support and preferably strengthen existing power structures against which the action is a threat. The reaction, therefore, is a mobilisation of defence, an act of life-support for a power whose position is disputed as a result of the emergence of a rival power; a prime example of this is when the dominant class of a dying paradigm defends itself against a new elite which has the winds of history in its sails. If we transfer this conceptual pairing to sociology we get trend and countertrend. A trend in this sense is a movement in time which is connected to the struggles of a certain group to achieve and manifest a social identity. A trend is an original impulse in the sense that it is not principally a response to another movement; it encounters resistance only when it collides with other interests, which it always does sooner or later.

A trend can be recognised by two distinguishing characteristics: partly that it has everything to gain from increased dissemination of information, and partly that it is intimately connected to and benefited by the territorial gains made by new technology. This means that the development of the Internet itself, as well as the social developments which presuppose or are benefited by the existence of the Internet, such as globalisation and the bourgeoisie's loss of power and status to the netocracy, can be identified as genuine trends. A countertrend, on the other hand, can be defined as a reaction against such a trend. It does not stand for anything other than the defence of the status quo, or a return to a past which has been highly sentimentalised. A countertrend is therefore principally concerned with combating and disarming a trend. Current examples of countertrends are the hypernationalism and isolationism which are being nurtured in various places in the western world, or so-called Muslim fundamentalism in the Arab world. These are both reactions against the dominant trend which is moving towards globalisation, secularisation and increased plurarchisation in society. A countertrend is always secondary in that it is always dependent upon a trend for its existence.

If we look at developments over a longer period, the pattern becomes considerably more complicated. Often one and the same social movement contains elements of both trend and countertrend. A good example is the environmental movement, which contains a "trendy" faction, which regards technological advances and open debate positively, and which claims that the only way to create an environmentally aware society in the long term is to use information and discussion to stimulate new research directed towards a new, green form of advanced technology. But the environmental movement also contains a "countertrendy" faction which is opposed to economic growth and wants to suppress information by forbidding research, and which more or less wants to enforce the relocation of city-dwellers to the countryside, thereby returning to its romanticised image of the past.

Another example is the international movement against free trade, which has thrived as a result of a comprehensive network constructed with the help of the Internet, and which must therefore be regarded, at least formally, as a trend. But in essence this movement is driven by the ambition to protect well-organised vested interests in wealthy countries against competition from cheaper imported goods, effectively shutting out poor countries from all profitable markets; this desire to limit information is clearly an example of a countertrend. For obvious reasons this sort of hybrid movement is extremely unstable. Their similarity to rapidly decaying chemical elements with large atomic cores has led us to define them as quantum-sociological phenomena.

Without making any qualitative comparison between trends and countertrends – Nietzsche for his part always favoured action – we can confirm that history tends to favour trends, and that countertrends are generally doomed to fail in the end. This is in the nature of things: a countertrend implies a limitation of access to information, a strategy which in the long run can never be anything but unsustainable wishful thinking. Countertrends play on people's fears and apprehensions about trends, but can ultimately only delay the historical process, never change it. Once information has begun to seep out into society the countertrend loses its potency, and the trend breaks through on all levels. However we must not forget that today's countertrend is often yesterday's radical and innovative trend; in tune with the times when times were different.

Nationalism and democracy were originally fine examples of social and political trends, completely in tune with the technology of the age and supported by an increase in information. But the forces which are struggling to maintain old institutions today are the countertrends of our time. In spite of the occasional success, these forces are doomed to failure in a broader spectrum. The Internet is a fact. Mass-communications are rapidly becoming interactive. On a purely national level there is little that is of any real political interest. This means that globalisation, secularisation and pluralisation are ultimately unstoppable.

At a generous estimation, bourgeois democracy is little more than 200 years old, but in that time an impressively tenacious myth has developed around it, and has convincingly maintained that its structure and ideology are eternally valid. This myth has been supported by the fact that bourgeois democracy, seen as a trend, was spectacularly successful in its heyday, and has contributed to an unparalleled increase in general welfare. In the mild delirium of the end of the capitalist era, bourgeois democracy has been seen, particularly after the pathetic collapse of communism in eastern Europe, as not merely a precondition but almost a guarantee of economic growth, despite the fact that it is easy to refute this rather unsophisticated thesis. If we turn our critical gaze towards Asia, for example, we can see that it is hardly thanks to democratic virtue that Singapore, a shamelessly repressive one-party state, has been able to demonstrate an incomparably strong economy for several decades, and that democracy has not paved the way for an economic miracle in Bangladesh. But in spite of this the myth lives on.

The American political scientist Francis Fukuyama has proposed the idea that bourgeois democracy represents "the end of history" in a Hegelian sense, a completion of the political process. History should be seen as process of development common to all people, a lengthy struggle between different social systems and philosophies, a conflict that will rage until all available possibilities have been tried and all unsatisfactory solutions discarded. Fukuyama is suggesting that all possible alternatives to bourgeois democracy have been swept aside. The decisive factors in favour of democracy are that it was best suited to cope with all the contradictory interests which arise in an advanced, global market-economy, and also that democracy, with its theoretical equality, was the best able of all imaginable alternatives to satisfy its citizens' need for recognition and respect. Fukuyama is both right and wrong about this. Bourgeois democracy is admittedly the most suitable social system in most respects, but only within the framework of the capitalist paradigm. When circumstances change, history is in motion once more.

So much ideological capital has been invested in the idea of bourgeois democracy that every attempt to examine its propaganda critically, and every discussion of possible faults in the democratic theory, have been rejected with distaste and labelled as heresy by the immense forces guarding this myth. A cult has been constructed around democracy and its rituals, a cult with strikingly irrational elements, a mass movement whose cohesive idea is the notion that the magical power of the correct political procedure can solve, or at least ameliorate, all serious social problems. The forms of democracy themselves are believed to conjure up sound political content. The correct method of taking decisions always results in wise decisions, according to this most potent of democracy's dogmas. The self-appointed representatives of political goodness have persuaded themselves, and anyone who has cared to listen, that there is no difference between one and the other, and whenever someone has the bad manners to point out that the Nazis, to take one striking example, came to power as a result of democracy and parliamentary elections, the response is always a series of diffuse explanations about the German population's "democratic immaturity", and the existence of that specific situation at that particular time, but seldom or never a recognition that democracy itself is no guarantee for good and wise decision-making.

Every exception is turned around so that it confirms the general rule: as soon as the democratic maturity of the population has reached a certain acceptable level, democracy will automatically produce wise decisions like a well-oiled political machine. This faith in the process of democratic maturity of the population is as unshakeable as it is unfounded. It is a matter of religious faith rather than empirical knowledge. According to the axiom, voters cannot vote wrongly, and if the people should nonetheless do so– like the Austrian people did when a populist right-wing party was elevated to government in the dying weeks of the 20th century, a result which the rest of the European Union strongly opposed – then it is simply a matter of pronouncing loudly that a mistake has been made and bringing about a correction through suitable actions.

This is all about the bourgeoisie parading democracy as if it in itself were something holy, yet being unable to accept democracy when it conflicts with the bourgeoisie's own interests. There is a considerable amount of hypocrisy in the ceremonial rhetoric of democracy. What is at best a problematic and fragile compromise between conflicting political forces, and at worst a method of legitimising the decisions that the elite wants to force through, is elevated as something good, eternal, and beyond doubt, something in accordance with the laws of nature. Feudalism maintained that people could believe what they wanted to so long as they believed in God; capitalist tolerance stretches to the point that people can vote for whomsoever they want to so long as they vote for the bourgeois democratic state in one of its sanctioned variations.

As is the case with all the obvious contradictions in a governing paradigm, this one is a sign that the old paradigm has played out its role. Just one decade after the discussion of bourgeois democracy as the culmination of history there is now, ironically, a growing consciousness of a crisis of democracy. Apathy towards utopia is accelerating, and democracy is being gradually stripped of both internal and external components. At the same time as technological development is promoting an entirely new ecology for humanity, the value of the democratic "recognition" upon which Fukuyama based large parts of his argument is being devalued. Fukuyama himself was the first to address the problems in his own cheerily optimistic theory with a dose of good old Nietzschean doubt. What is the point, he wondered, of universal recognition in the end? Who will be satisfied with something which is available to everyone? Invoking Nietzsche, Fukuyama pointed out the connection between the tolerance on offer and a dissolution of values that would be fatal for society. What genuine sense of self-respect can allow itself to be detached from every form of achievement? And is universal recognition actually possible? Can we ever hope to get away from a system of values and a hierarchy of various achievements?

An entirely equal person is incapable of contemplating social and moral issues, because to do so presupposes a basic difference between good and bad, which is at odds with the extreme tolerance which is the hallmark of "political correctness", the last desperate attempt by an imploding democracy to flirt with everyone at the same time. Creative momentum is lost when society values everything as equal and has as its aim the equal distribution of resources. Stagnation accompanies stability, which leads to a gradual corrosion of democracy, both in its inert and active states. But stability ought in this context to be the least of our problems; it is a chimera, a socio-scientific abstraction. Stability does not exist in complex systems; Fukuyama's great mistake was to confuse the model with the reality.

Democracy has become a great worry: it demands attention in the form of education and nurture, which is why all sorts of authorities and research institutes are insisting on comprehensive budget increases. The warning implied by this is that we can no longer take democracy for granted. Democracy must at all costs be strengthened, runs the message from those whose power is entirely dependent upon democracy, and who for obvious reasons have no interest in any form of plurarchy. Propaganda declares that the only alternative to bourgeois democracy is a nightmarish dictatorship which, as history has illustrated, is all too easily attainable through exemplary democratic means, while plurarchy is suppressed as best it can be. There is an impression that it does not exist, that it is not even conceivable. This is why it is highly ironic that the Internet is being promoted as the tool which will facilitate the ultimate triumph of democracy. In actual fact the Internet is responsible for the new media-technological ecology in which plurarchy is thriving on natural selection and in which democracy is therefore doomed to failure. The crisis in democracy is here to stay. From now on democracy will be synonymous with the crisis affecting both itself and the whole of the declining paradigm. As a result of the paradigm shift the unthinkable has become thinkable.

Informationalism will force the development of new political structures. One concrete political problem is that in areas where politicians want to make decisions there is no longer anything to decide about. The market and the economy have moved. In this situation growing federalism will be the final straw for democracy, because developments are moving ever more rapidly towards a global state. The introduction of such a project is probable, not least because the survival of the political class is at stake. A world trade organisation will be followed by a world tax organisation, and so on. But all efforts in this direction will be in vain, not because any attempt to mobilise enthusiasm for the project amongst all the world's voters would take immense effort, but because plurarchy will already be an accomplished fact before a detailed plan for a global state has even been sketched out. In the virtual world, politics will become powerless.

But even the virtual world will spontaneously establish certain global state structures. The electronic elite will establish a new lingua franca, a Net-Latin based upon English, already the global language of communication for the netocracy. This highly modified form of English, in which subcultural dialects will come to the fore, where standard phrases will be drastically shortened, where innovative neologisms will be encouraged and sub-clauses abandoned, will be the universal language of the global networks.

The overwhelming Anglo-Saxon domination of the international music and entertainment industries indicates the central role played by a common language in the current medialisation of the world. We can already see how the English-speaking countries of various continents have established a lead over their neighbours in the race towards informational society. This is obviously the case with Great Britain and Ireland, Canada, Australia and South Africa, all of whom are playing leading roles in spreading the Anglo-Saxon-dominated Internet culture to every corner of the world; this without even mentioning the role of the USA. But even Hong Kong, Singapore and India are doing the same in Asia. Or the Netherlands and the Scandinavian countries in Europe, where proficiency in English has become so advanced that it has practically become a second native language. The convergence of global communication is leading to an increasing demand for some form of uniform, common language, which will further benefit the Anglo-Saxon media industries which were already well-developed before informationalism, and which are now the main suppliers of content to the Internet.

It is entirely logical that it has been easier for small countries with highly restricted domestic markets to think and act globally than for countries with large domestic markets. The relative uselessness of their native languages has forced them to adapt and gradually switch to the English-based Net-Latin. Those countries which have had ambitions of competing with English as the global lingua franca of the Internet – primarily those countries in which French, Spanish, German, Arabic, Japanese and Chinese are spoken – are experiencing, on the other hand, comparatively large and intractable problems in connecting their national network structures to the global net. For this reason there will be initially a limited number of people from these linguistic areas taking part in the international networks. While we are still waiting for comprehensive translation software, language is still a barrier.

Local and regional languages and dialects are maintaining their strong position within large populations, but are gradually assuming an recognisably inferior position. Native languages will live on among the consumtariat, while the netocracy will regard them as a kind of amusing hobby with sentimental connections to the past, one of many sources of entertainment and amusement based upon identity. In much the same way as the middle-class tourists of former times spouted sentences culled from phrase-books during their holidays abroad, the netocrats will amuse themselves during their summer holidays in their digitally wired country houses by practising the characteristic traits of the local language, all with a high degree of ironic distance. Within the porous framework of Net-Latin there will of course be a number of different dialects, but these will not be geographically anchored but instead linked to different virtual subcultures, and will function mainly as identification markers to distinguish their speakers from the members of other electronic tribes.

On important reason why the USA has found it so easy to adapt to virtual culture is the pre-pluralistic attitude to geographical space which has always suffused the American mentality since the time of the first settlers. Majority decision-making has never had the same strong roots in the collective consciousness of North America as it did in Europe, thanks to the concept of the "frontier", the continual exploration of the boundary of the unknown. When an American did not see eye to eye with his neighbour, there was always the possibility of upping sticks, moving further west and breaking new ground. There was a lot of spare land, and colonisation took a long time. When the settlers had reached the Pacific, they continued to Hawaii and Alaska. Connections to secure territory have never been particularly strong in American culture. There has always been an emotive and irresistible desire for the unknown, a permanent preparedness to move on, a romanticisation of the nomadic. We need to bear this in mind when we look at the great space projects of the 1960s and 70s; once the earth had been mapped and colonised there was only space left. Space became "The Final Frontier".

But only as long as our thinking was confined to physical space. The truly revolutionary adventure was waiting within the digital universe of satellite-linkups and fibre-optic cables. During the 1980s the Internet was made available to everyone, a completely new "frontier", and the first people who began to explore this virtual world were precisely the outsiders of American society, lone wolves or members of diverse subcultures who were ill at ease in the dominant cultural climate. They felt out of place, so they upped sticks, turned their back on mainstream culture and moved on, breaking new ground in a new world. This virtual mass migration, described by the American social theoretician Mark Pesce as "the Gnostic Frontier", marked the beginning of colonisation of a new world which was quite literally infinite. A continuous stream of new and unexplored virtual territories awaits the nomadic Americans. The journey need never end. The pre-pluralist tradition has prepared the way for a smooth transition from democracy to plurarchy, and the shift will be experienced as relatively mundane. What will be harder for Americans to accept will be the devaluation of money and the death of capitalism.

Another sacred social institution which is undergoing a fatal and intensely debated crisis at the onset of the informational paradigm is the nuclear family. There is a good deal of confusion in this discussion, in which conservative elements are arguing for a return to "tradition". What we need to be clear about is that what is usually called the "traditional family", where the man goes out into the world to earn money while the woman stays at home and looks after the house, is anything but traditional. In actual fact, the "traditional family" was a child of the 1950s, a result of increased welfare and the dream of fully-fledged suburban bliss free from social obligations. The middle-classes moved out to the suburbs to get away from an urban community that felt altogether too stifling and oppressive. The suburbs, with their lush vegetation, were attractive because they appealed to our tribal instincts.

The real traditional family is something entirely different: an economic union and socialisation project which links various generations together. The nuclear family – which is, interestingly enough, the same age as another post-war phenomenon, nuclear weapons – ought therefore to be regarded as an integral part of latter-day capitalism's obsession with individualism: total freedom at the cost of total isolation, consumption instead of communication. There is a simple and obvious logic to this. It is in the interests of the state, as well as of capital, (once again in the same boat) that people develop into "independent individuals" rather than as members of wide-ranging networks of social communities. Quite simply, it makes it easier to exercise central control, and at the same time people develop a sense of individual responsibility for their own self-realisation, which is expressed in intense and often therapeutic consumption.

The nuclear family arose because it was the smallest, most individualised social entity that was feasible once the necessary requirements of reproduction had been taken into consideration. But as far as the state, and capital, are concerned, it does not need to be stable at all. A single parent is more dependent upon subsidies than a couple, and is therefore more submissive and more easily controlled, and a marked increase in the number of single-occupancy homes also means, naturally, that consumption increases dramatically. The ideal would be for everyone to live alone in their own home, with their own car, so that optimal demand for houses, cars, sofas, cookers, etc, could be achieved. The path of the isolated, independent individual towards self-realisation always passes through increased consumption.

For these reasons the great wave of divorces that has swept the western world since the mid-1960s does not actually constitute a violation of capitalist values; on the contrary, it is a logical continuation of increasing individualisation on all levels, and of the escalating development of social structures under the capitalist paradigm: from village communities and the tribal family which consisted of several generations, to the single urban citizen, who himself constitutes the ultimate capitalist family unit. This will not prevent a number of noticeable elements of the introductory phase of informationalism from strengthening some of these tendencies. The clearest example is when industrial production decreases to the benefit of information management and an expanding service sector. This means that a large number of poorly educated men will be rendered superfluous in the labour market. At the same time, an increasing number of career opportunities for women are opening up, suggesting a further increase in divorce-rates and the number of single people.

Sexuality is another matter: the Pill meant that sex could be differentiated from familial responsibility. Francis Fukuyama has suggested that the Pill not only granted women the freedom to live out their desires without consequences, but that, primarily, it released men from any sense of responsibility for the children that were born anyway. Maternal care for infants is highly biologically programmed, whereas paternal care is more of a cultural product, and therefore more sensitive to disruption. Added to this is the fact that reproduction is gradually becoming separated from sexuality generally, with both fertilisation and pregnancy being increasingly taken over by biotechnical laboratories. This is hardly going to strengthen family bonds.

The freedom from censorship offered by the Internet also means that sexuality is becoming less dramatic, and is just one pastime amongst others: the link between sex and cohabitation is being weakened, which means partly that sexuality is being distanced from relationships generally, and partly that a wealth of cohabitation forms is appearing in which sex plays no part. You no longer need to live with your sexual partner, just as little as you need to live with your tennis partner: You do not need to have sex with the person you are living with, as little as you need to have sex with your boss or your therapist. Conventional forms of personal relations are dissolving and disappearing. Circumstances will determine the form of a relationship, and not vice versa.

The lack of a normative majority on the Internet means that every conceivable sexual preference is becoming socially acceptable; every taste has its own more or less global network. Homosexuality, sadomasochism, different forms of asexuality: everything is going on alongside more generally inclusive activities, which is providing inspiration for the most disparate lifestyles and forms of relationship. The Internet has already stimulated unparalleled experimentation in areas of sexuality and relationships, a development which is now exploding. The keyword in this context is queer. Queer culture implies the liberation of heterosexual society from the compulsion to be normal and normative, and is growing rapidly through imitation of the successful networking of gay society. Sexuality is no longer a subject for control and regulation, is not made to perform a socially cohesive role within the State or the Market, but is now the basis for the creation of new, more or less temporary, tribal identities on the Net. Queer culture is thus a genuine trend. Sex and cohabitation are being replaced by sex and tribal identity.

It is the search for tribal identity within the practically limitless framework of virtual nomadic society which is forming new, fluid family structures. This tribalisation of culture, this virtual return to nomadic existence made possible by electronic networks, means that permanent homes are no longer regarded as the fixed point in life, which itself stimulates further unconditional experimentation with lifestyles. The rapid increase in mobility and speed in society is leading, for better or worse, to a growing sense of rootlessness. The new sense of homelessness is simultaneously imposed and desired, a burden and a possibility. The new nomadic life implies a permanent migration between different cities, different workplaces, different identities.

The mobility of the netocracy, first virtual, but increasingly physical, is leaving deep traces on society and culture. The idea of a home and a place of residence is changing beyond recognition. The higher your status, the greater your degree of mobility. The consumtariat, securely settled in fixed dwellings, is easily accessible to the homogenising effects of the mass-media, and the remaining political powers and their ever more desperate tax authorities. The netocracy is abandoning the traditional suburban fortress established by and for nuclear families. Now hotels, monasteries and meditation centres are the major influences on the new elite's way of life. The function of the secure home to act as an identifying factor is disappearing and being transferred to the Net, to the virtual equivalent of the home: the homepage. A homepage is what a genuine Netocrat can accept as his fixed point in existence. As long as it can be regularly updated!

Another institution in deep crisis is public education, one of the fruits of large-scale population growth in Europe and North America during the 1800s, which in turn, like the explosive population growth of the Third World during the 1900s, was a consequence of the habitual delay between technological changes and cultural behaviour patterns. As we have earlier asserted in our analysis of mobilistic diagrams, history moves far faster than people can possibly react. Changes which affect material circumstances often achieve their cultural impact only after a delay of several generations. Populations grew in Europe and North America because infant mortality dropped and the general preconditions of life improved radically, at the same time as the birth rate remained at its earlier high level. A noticeable decline in the birth rate did not occur for several decades, which meant that for a long time there were colossal numbers of children. The working-class families of the countryside and the city slums were turned into veritable production-lines for the making of children.

This development forced wide-reaching political activity: the growing mass of children needed to be cared for and protected in a satisfactory way, to be civilised and brought up as useful members of society, as diligent workers and insatiable consumers. These ambitions were co-ordinated in the middle of the 1800s in a new institution: the state school. Apart from their practical function, schools also had an ideological role. During the early years of industrialisation the hoards of children had been put to hard work in factories at an early age. But the middle- and upper-classes eventually forced through legislation regulating child labour, and in time also introduced compulsory education. Partly inspired by the romantic cult of the child, and its imagined closeness to some original state of innocence and purity, they wanted to protect and nurture the young, sheltering them from the more brutal sides of life and carefully introducing them to the secrets of adult life.

It is important to remember that childhood is a cultural product, created during the Renaissance. During the age of feudalism children were not regarded as belonging to a special category, and were not assumed to have any special needs, and their entry into adulthood was not a question of nurture and education. Children were purely and simply small people, far too weak to be of any real use. The word "child" did not describe an age, but a relationship; you were the child of so-and-so, and always would be, which underlines the obsession of feudal society with family names.

The advent of childhood during the Renaissance is, as Neil Postman has pointed out, closely connected to the printing press. This new information technology brought with it a new perception of what it meant to be an adult, the ability to read; consequently the definition of a child was the opposite: a person who had not learnt to read. One did not automatically become an adult with time, but through education. Education is in some respects a revolt against nature: a small child is made to sit still and fret over the alphabet and other studies when play and other physical activities are much more appealing. In this way education coincides with the need for the child to control its impulses. Childhood thus became one of the basic discoveries of capitalism, and as such became the subject of endless ideological conflicts.

Universal education meant that the process of becoming an adult and a member of society became industrialised, and encompassed even the children of the working-class. What was grandly described as a human right was at the same time a social duty. Someone who did not go to school could not become an adult, nor, therefore, a fully-fledged citizen. Society's values were instilled, and the necessary capabilities to ensure that society's needs were fulfilled were taught. Education itself was a visible example of progress, and the idea of progress suffused the whole enterprise. Everyone had reason to greet universal education warmly: for the working-class it was necessary if its children were to have even a theoretical chance of climbing socially, and for the bourgeoisie it offered the chance to recruit new talent to the administration of the state apparatus, and to form the remainder of the working-class into an effective workforce for the factories.

Nothing was left to chance. The whole enterprise was rigorously planned according to the most advanced pedagogical programmes of the time, and the model for the organisation was taken from related institutions such as the military, and mental hospitals. Schools became the instruments of selection for the capitalist meritocracy, and functioned satisfactorily so long as the labour market demanded clearly defined competencies for a more or less stable list of standardised careers: in other words, as long as there was a direct link between education and working life. But with the breakthrough of informationalism the entire capitalist notion of the "career" is crumbling. A crisis within our schools is inevitable.

The informational labour market has an entirely new structure. Employment is no longer a lifelong contract, and length of service is no longer of prime importance. Business organisations are becoming less rigid and are concentrating on temporary projects, for which people with specific competencies are employed. Temporary constellations are created only to be dissolved when a project is completed. Education is never a completed chapter but must be constantly updated. Every new task involves a new situation, which generally requires new knowledge. The unavoidable consequence of this is that all diplomas, titles and certificates are practically worthless the day after the exam. This, in turn, means that schools are really losing every other role except for those of a holding-pen and a place for social training; children can learn more sitting in front of their computers at home than at their desks at school. In an increasingly fragmented and changing society, the whole idea of centralised, homogenised schooling linked to the nation state seems outdated.

Already, increasing demand for further education and the development of skills both inside and outside of business has led to many old academic institutions sensing that change is in the air. Politicians fight to be able to inaugurate new subsidised business parks aimed at growth industries like information technology and bio-technology, connected to universities and colleges. Increasingly costly educational packages are being tailor-made for companies in a seemingly limitless spirit of generosity. But it would be a mistake to take this as a sign that traditional educational institutions will play a leading role in informational society. Business parks are not a trend but a countertrend, an increasingly desperate attempt to protect old hierarchical structures under new circumstances, and are therefore doomed to fail. These actors from the old paradigm are far too closely tied to their old historical positions to be able to move easily and quickly enough to survive in the virtual ecosystem.

The feudal roots of academic culture are evident in its hidebound fascination for titles and old qualifications. Its closed frame of reference, its rigid hierarchy, its incapacity to assimilate criticism as something constructive: all of this is creating a problem of credibility. The academic world appears, in the eyes of the netocracy (whose scepticism towards self-proclaimed experts in every field is practically constitutional), to be obsolete and corrupt. This opposition between the netocracy and the academic world is to a certain extent superficial. But beneath the surface there is a more fundamental difference in attitude which is becoming more apparent as the universities desperately seek to protect their increasingly exposed position in informational society.

To put it simply, it is a question of opposition between two completely different temperaments. For the netocracy speed and an overview are the primary requirements, whereas traditional research prioritises thoroughness and depth, which explains the persistent studies of a stability that is purely fictional: purely theoretical constructions with little connection to reality, which dominated social science during the 1900s. The netocracy is interested in change, whereas academics are concerned with static models. Or, to put it another way: the netocracy is interested in bodies in motion, while academics prefer to perform autopsies on old corpses. There will be not so much an encounter as a brutal cultural clash between these two participants, a clash which can only end badly for academics, whose obsessively neurotic attachment to scientific scrupulousness, references, footnotes, etc., make them incapable of attaining the speed and overview which appeal to the netocracy.

What the netocracy is seeking and needs is something quite different to what the universities are offering: the ability to absorb and assimilate large amounts of information, combined with an intuitive understanding of what is relevant in each specific situation; quick associations and irrational playfulness rather than conscientious analysis of sources. The netocracy's attitude to knowledge is at the same time instrumental and aesthetic. When the netocrats do not find what they want in either universities or business parks, they will turn their backs on the academic world and construct their own primarily virtual institutes and think-tanks, free from outsized administration, intellectual snobbery and a tyranny of detail.

Education will in the future be characterised by an interactive and constantly adapting pragmatism. It will be offered and developed on the Net in the form of small, precisely adapted modules, specially designed for the task at hand. It will be the student who decides the rules, not the institution. There will be stiff competition between these sophisticated and flexible systems, and in this market there will not be room for capitalism's inflexible and resource-hungry monster universities. There will no longer be any need to sit isolated in some regional college, divorced from the labour market, adapting to obsolete conventions in the hope of getting a diploma which, like the Soviet Order of Lenin today, only has kitsch and curiosity value. The future of collective education is behind it.

This development is being hastened by the academic institutions' inability to create functional networks. The mailing-lists and web-archives that have been launched so far by the leading universities in Europe and North America are based upon the entirely mistaken belief that creativity and problem-solving can be stimulated with the help of eternally rumbling debates, open to everyone. This attitude seems unforgivably naïve to a netocratic observer. Within the netocracy there is an acute awareness that qualified networks can only function if they are created as time-saving arenas for contact between selected participants, and as contact points for the exchange of exclusive information. This requires strict and sensitive curators; it requires the ruthless manipulation of information, presentation and accessibility. This means, in effect, that practically any teenager is capable of managing his own private network more effectively and, above all, more purposefully, than naïve and clumsy universities with their feeble presence on the Net, in spite of all their enormous subsidies.

It is ironic in this respect that the Internet was originally created for, and partially by academic institutions that did not understand how to use the new medium with whose management they were entrusted. Recent developments have left them behind. We can see once again how creativity rather than financial resources or politically directed regulations is decisive in the allocation of power in informational society. It follows from this that there is no reason for the netocracy to boast of academic titles; on the contrary, it will be more prestigious to highlight a lack of formal qualifications. In the eyes of the netocracy, completed educational programmes and doctoral titles are not signs of merit, but an indication of an inexcusable lack of judgement. Universities will come to be regarded as protected workshops for intellectual therapy, and anyone who has spent his time there will be treated with an increasing amount of suspicion. At the same time, academic institutions represent a powerful interest which cannot be disregarded; they will be perceived as a potential source of countertrends which might harm the netocracy. For this reason it is unlikely that the netocracy will be content merely to ignore the academic world, but will actively oppose it, if only by excluding its representatives from attractive networks.

Trends are encountering countertrends on every social and cultural level in the transition between capitalism and informationalism. One far-reaching trend, related to welfare and education, which is showing signs of getting stronger, is the diminishing birth-rate of western countries. Giving birth is simply not fashionable any more. In a majority of western nations women are giving birth to less than two children on average, and the consequences of this are not hard to work out: the population is declining at an increasing rate. The politicians of the 21st century have no hoards of children to take care of; one of the truly large problems will be the exact opposite – how to manage the decrease in population numbers. A shrinking number of young people is expected to provide for an increasing number of old, and steadily older, people. In a democratic society the elderly would be able to force through a gerontocracy by using their majority position to outvote the smaller number of young people. But, as we have already seen, in a pluralist society there is no connection between positions of power and purely numerical superiority, as little as there is any direct connection between power and money. On the contrary, it is probable that a minority of young people will have power thanks to their relevant abilities and their greater manoeuvrability on the Net.

While the netocrats are experimenting with identities and lifestyles in their heavily guarded networks, the consumtariat is held in place thanks to the Disneyfication of the whole of the popular cultural landscape. Entertainment, consumption and leisure time are melting together into a single, enormous industrial sector. Large holiday resorts with adjacent entertainment factories are being built near airports, where the consumtariat are carried to be entertained to sleep. The very latest and most expensive entertainment technology is offered by so-called "multi-media theme-parks": collective experiences for alienated keyboard-slaves. On the innumerable recreation sites on the Net there are interactive soap-operas and every conceivable variety of bingo, lotto, betting, and every possible sort of game. Each form of game will have its own television channel and a host of homepages. This theatre performs all day, every day, always.

It will become considerably more difficult than before to use mass media for propaganda purposes, as we have already discussed in an earlier chapter. It will not be possible to use the same clumsy strategies as during the days of centralised one-way communication. Simultaneously, increasingly medialised, virtual reality will become sensitive to media manipulation. Any attempt to use the media will increasingly be an invasion of reality itself. The boundary between one and the other will become steadily less distinct, and will eventually prove more or less impossible to maintain. As a result, propaganda will become invisible, no longer discernible even to an expert; it will become its own reality, in its own right, the ever more subtle manifestation of the power of the elite in the form of pleasant and soothing mental massage for the masses.

Both art and philosophy are trying to find new tasks and new means of expression. The ambition to create an all-encompassing synthesis, total art, a universal explanatory model, has gone now. Virtual daily life will make that sort of thing seem as thoughtless as it is pointless. The linguistic philosophy of the 20th century has left us with an awareness of the limitations of our conceptual apparatus. The big problem is that the amount of information on offer is increasing exponentially, while our perception and our capacity to deal with incoming impulses is developing with the studied slowness of biological evolution; in other words, scarcely noticeably at all. The virtual world is rushing away from us, that is the tragic realisation of mobilism, which presents us with the task of making what in reality is an unmanageable world somehow manageable. This necessarily artificial level of comprehension contains an increasing amount of incomprehension; we are becoming increasingly dependent upon our ability to create functional models in order to orientate ourselves.

The new rationalism is therefore becoming transrationalistic, and contains a fundamental understanding of the unavoidable limitations of rational thought; it denies every form of transcendentalism and metaphysics at the same time as humbly acknowledging the shortcomings of rationalism. A mobilistic credo would be able to take as its starting point Nietzsche's exhortation to capitulate willingly to infinity, to be filled with "the joy of tragedy". Or Spinoza's exhortation to love this unfathomable world despite the fact that we can only expect frosty indifference in return. We must, according to the mobilists, abandon our infantile need for response and affirmation. Quite simply, we have no choice, life cannot be anything other than what it is.

The vacuum which remains when rationalism has given up, the "trans-" in transrationalism, can only be filled with painting, literature, music and all the new hybrid forms of art which are being opened up by new technology. Creative possibilities are practically limitless. The flip-side of the coin is that art, more than ever before, will become the exclusive province of specific electronic tribes. Art will probably not reach other groups to any noticeable extent, partly because of the extremely targeted output of media on the Net, and partly because of other groups' lack of a frame of reference. The whole apparatus which makes slightly more demanding culture comprehensible exists only under highly fragile circumstances on the Net. This is why culture will become yet another dividing barrier separating different groups from one another in the electronic class society, and a cohesive and identity-supporting factor only within narrowly-defined groups.

Informational society is anything but equal. And the inequality it offers seems more "natural" than was the case in earlier times because its meritocratic element is so large, because power is so difficult to localise, and because its representative mechanisms are so discrete. The netocracy is fairly untouchable; it has not taken anything from anyone, and its position of strength and status is built upon its undeniable suitability to thrive in the new ecosystem created by information technology. Nor does the new underclass share the same exciting, sexy attributes and the justified pathos which the underclass of the capitalist paradigm could demonstrate, and which aroused a certain level of sympathy. The consumtariat is the underclass because of its own lack of social intelligence, according to the norms of intelligence established by informational society.

No doors have been closed to anyone; the problem is that it requires a special talent to understand how to grasp the handle and get in, a talent lacking in the masses. Is this inequality necessarily unfair? And if it is, according to whose criteria? And, if so, what can be done about it? Ought we to hold back those who have the ability to make the most of the opportunities on offer? Ought we to carry on giving new chances to people who have failed so many times before? How can we solve the problem of increasing inequality in a society where inequality cannot be rectified by redistribution? We have not yet worked out how to swap brains with one another.

11.
BEHIND THE FIREWALLS – NETOCRATIC CIVIL WAR AND VIRTUAL REVOLUTIONARIES

As the central capitalist institutions collapse, an echoing vacuum is appearing. As long as they retained any authority, these institutions fulfilled a stabilising function; what is coming instead will be a state of institutionalised turbulence whose dynamic is extremely difficult to predict. Certain trends are abundantly clear, but when so many trends and countertrends interact, there are so many parameters and the level of abstraction is so high that even the most sophisticated guesses become a sort of intellectual meteorology: reliable only at very short range. Despite this, there is every reason to gather such qualified prognoses as are possible, most appropriately by trying to identify and analyse the social tensions that will characterise informational society.

Political and cultural debate within informational society will take place against a background of entirely new circumstances. The debate about equality that took place under capitalism will appear hopelessly tied to a bygone age, where status and power were primarily distributed according to an arbitrary system. Neither family background, wealth, gender nor skin-colour will have any decisive significance in informational society, where individual status and power will instead depend upon the individual's capacity to acquire and manage information, upon social intelligence, receptivity and flexibility. The liberal ideal of equality: equal opportunities for everyone to realise their life's project, has therefore already been realised in practice (while the socialist ideal of equality: equal rewards for all regardless of context, must be regarded as discredited and consigned to history with the collapse of the communist utopia).

But at the very moment this realisation takes hold, it will be clear that informational society is in several respects more unequal and static than any other. What we call "the new sociology" is devoted to describing these conditions. The system is characterised by great permeability on all social levels, and consequently great mobility for the individual, but these porous structures are, in return, much stronger. The mechanisms of meritocratic classification will become increasingly refined, each and everyone with enough talent and initiative to constitute a threat will automatically be promoted to a privileged position within the network hierarchy and incorporated into the elite. And it is difficult to imagine any political activism aimed at the inequality of the brain, or the fact that talent is rewarded. It is, after all, only "natural"!

The new patterns of this class division under informationalism will contribute, together with network society's increasing opaqueness and the collapse of traditional left-wing ideology, to creating a seedbed for a dramatic increase of violence in society. The consumtarian protest movement will suffer a chronic lack of leaders – because potential talents are constantly absorbed into the netocracy – and will have little ideological sophistication. Its thinking will be contradictory, its actions erratically sporadic and impulsive. Social discontent will be blind. Consumtarian rebels will lack the old workers' movement's education and discipline, and will have no long-term objective. They will have no ambition to unite the consumtariat around a common cause, either within or outside the system, and no-one will believe in either organised revolution or revisionism: a sort of gradual netocratisation of the consumtariat through political struggle and hard work. What remains will be a sort of revolutionary aesthetic, a romanticisation of resistance as such, an intoxication of spontaneous, confused, collective destructiveness. But that will be all.

In this context it is important to remember that the consumtariat, in contrast to the old working class, lacks any solid conviction of a brighter future. A consumtarian rebel could never enflame his colleagues by claiming that the future belongs to the underclass; there is thus no notion that violent expressions of discontent would be progressive in any sense whatever. Instead, the consumtarian rebel is more likely to flaunt his regressivity and his hatred of both present and future. The consumtariat's revolutionaries will therefore have no ideological connection to either the old workers' and trade union organisations, or the peasant revolts of feudalism. They will take from these precursors at most only their rhetoric. Consumtarian rebels will instead take their ideological inspiration from the closed guilds of the Middle Ages and the puritan revivalist movement of the Enlightenment; a desire for isolation from their surroundings in expectation of the end of time and the collapse of the universe.

Early precursors to these consumtarian protest movements are already emerging in the transition between capitalism and informationalism, in the form of various headline-grabbing doomsday sects. These sects recruit their members from and are most attractive to the underclass of mass-medial society, which incorporates both the last remnants of the traditional working class and the expanding consumtariat. Doomsday sects are not a geographically localised phenomenon, and are appearing in disparate parts of the world – some that have attracted attention have been based in the USA, Switzerland, Japan, Russia and Uganda – which means that their appearance cannot be explained with reference to specific national cultures. They are, instead, a global phenomenon, an early example of consumtarian counter-culture. While the ambition of the netocracy is to conquer the world, these groups are turning their backs on hostile surroundings and are willing to cause damage before it is time to meet the group's own, self-chosen, physical destruction.

This distanced attitude towards their surroundings, this construction of a parallel reality is something which seems quite natural to the citizens of an increasingly medialised society, where the boundaries between the "reality" that was so carefully protected under capitalism and the fantasy promulgated by the media are increasingly difficult to discern, and progressively less interesting to maintain. The news is entertainment, directed and presented according to the aesthetic of entertainment; politics has, in the words of talk-show host Jay Leno, become "show business for ugly people", a sort of tv-drama about sensationalised social problems. Supply is usurping demand thanks to increased welfare and refined advertising; trademarks give a product an admittedly fictitious but no less powerful personality in an economy where entertainment is a central value; lifestyle is replacing life. This development is being strengthened by the fact that the netocracy is consciously turning its back on "reality" and taking refuge in its electronic tribes.

The arrival of informationalism and the breakthrough of the interactive media constitute yet another great step towards what Jean Baudrillard has called "hyper-reality". There is every reason for a lot of people within both the netocracy and the consumtariat to prefer fiction to reality, because the former allows far greater choice when it comes to the construction of a social identity. There is a gradually increasing Disneyfication of our entire environment; old ruined castles are renovated and turned into places for stressed city-dwellers to go on outings, unprofitable farms are becoming theme-parks with an agricultural theme (so-called "agritainment"); cruise ships, hotels and entire destinations are being planned for longer or shorter stays in carefully realised fantasies, and so on. People are, to a great extent, becoming actors in their own lives, playing the "role" of themselves more or less convincingly. Reality is becoming an ever more subordinate part of hyper-reality, just as nature is becoming an ever more subordinate part of culture. There is no longer any actual reality, just virtual arenas in which performances are staged. The virtual environment is therefore becoming entirely synonymous with "the environment".

By the end of the 1990s, the Internet had reached distant villages in India and Latin America where there was still no running water. The netocracy has grand colonial ambitions, which is why the consumtariat need not worry about not getting access to exciting new technology and all it offers – quite the reverse! The underclass's only real chance to express discontent with its subordinate position will be to refuse to take part in the role-play of informational society. The aesthetic of passive resistance will then become a self-selected act of exclusion, while the strategy of active resistance will be violent demonstrations, inspired by the Luddites of early industrialisation who smashed the machinery that was undermining the value of their manual skills and destroying the preconditions for their traditional way of life. The threat of violence is the only thing that will make the dominant class listen.

Consumtarian rebels will therefore establish reactionary cells for the production of countertrends in an effort to achieve both technological and social exclusion. They will be following the path of the revivalist movement rather than the workers' demonstrations of 1st May, trying to break out rather than reform from within. Their answer to technology and sedative entertainment will be violent resistance. The effects will be dramatic, but this is not to say that the drama will be effective, since the consumtarian rebels will be far too lacking in resources. A genuine informationalist class-war will only be possible when consumtarian rebels get support from outside their own ranks, which will only happen when the only partially apparent unity within the netocracy splits at the seams. As a result, powerful, anti-netocratic networks will spring up for the first time: an unholy alliance of the consumtariat's revolutionary desperados and netocratic class-traitors, and it will be between these alternative, power-hungry hierarchies and the genuine netocracy that the informationalist class-struggle will be acted out in the form of irregular, explosive and potentially violent confrontations.

One inescapable precondition for this struggle is, then, internal netocratic conflict which is fundamentally ideological in form. This conflict is inherent right from the start, and concerns the infernally thorny question of immaterial rights. Just as the aristocracy once accepted its historical fate and co-operated with the bourgeoisie that was taking its power, so the bourgeoisie is now smoothing the way for the netocracy by helping to legalise the ownership of ideas. In order to understand the background to this virtual issue of ownership, it is necessary first to be familiar with four key concepts in this context: copyright, patents, encryption and firewalls. These four functions form the basis for what is called "the new economy", and therefore also the basis of the netocracy's appearance and assumption of power.

Copyright means the exclusive right to exploit, or control the exploitation of every form of immaterial right. A patent is the exclusive right, or control over the right, to exploit a certain invention over a certain limited period. Copyright and patents were fundamental functions even within the late-capitalist economy. With the breakthrough of informationalism, these rights apply, to an increasing extent, to digitally produced, stored and distributed information. The first digital products to be protected with the help of copyright and patents were software for computers and music stored on CDs, but with the growth of the informational economy, the value of digital information has grown phenomenally. This development means that the sale of ideas and design constitute a steadily growing proportion of the value of the economy as a whole.

This in turn means that the question of the copyright and patenting of digital products is becoming ever more central in informational society. It is a survival issue of the highest priority: the growth of an informational economy will be hampered and delayed if there is a lack of laws and rules in this area that are in tune with the times, and a judicial apparatus which is capable of implementing them. Enormous resources all over the world are therefore being expended on strengthening the legal protection of the right to exploit ideas. Legislation and police activities in different countries are being co-ordinated and standardised. One leads to the other: the fact that the informational proportion of the economy as a whole is constantly growing leads to an acceleration of these processes within business law, which in turn leads to increased informational growth. The growing netocracy and its allies, such as venture capitalists and specific political interest groups, have good reason to unite to protect their immaterial rights: their own survival is directly connected to the success of the project.

In countries like Russia, China, India and Argentina, there was no initial inclination to respect copyright and/or patents, because these judicial constructions only seemed to favour the already highly developed economies of western Europe, North America and Japan. Instead, these countries developed a strategy of imitating ideas and digital products which had originally been developed elsewhere. Without compensating the owners of the copyrights and patents, they mass-produced and sold cheap pirate copies of computer programmes, music and drugs, for example.

But a combination of heavy pressure from European, American and Japanese interests on the one hand, and the development of the countries in question in the direction of an informational economy on the other, has led to a radical U-turn on the issue. Globalisation has thus led to the political establishment all over the world uniting around the matter of protecting immaterial rights. The price exacted from anyone who chooses to remain outside this consensus is far too high, namely exclusion from the informational economy. For this reason, local authorities in China, Russia, and other places are cracking down on pirate-copying with considerably greater energy than before, at the same time as they are developing their own systems of copyright and patenting. Netocratic entrepreneurs and their capitalist investors can start to breathe easily again.

But within business and politics there is still not a complete understanding of what the Internet and technological development will mean. Culture on the net has its own dynamics, its own driving-forces, which are often at odds with late-capitalism's interest in the ownership of ideas. This means that the whole global system for copyright and patent agreements is starting to be eroded from beneath. On the Internet people all over the world can set out digital information on their homepages; this might be text, images, music, film, programmes, and so on. They can exchange this information between themselves as they choose, without any intermediaries or regulation, and without having to take into account any legal limitations or someone else's claims to copyright or patent. The people eagerly encouraging and carrying out this activity are netocratic class-traitors. The traffic in pirated information increased phenomenally during the 1990s. Neither national nor supra-national police organisations have any real chance of controlling or, still less, prosecuting this sort of activity, because it takes place in virtual space and therefore lacks any geographical basis.

The defenders of immaterial rights are also fighting in an increasing ideological headwind. A prohibition against copying physically tangible products – tables, cars, boats – is easy for almost everyone to accept. Someone who has constructed and manufactured a product also owns the right to control both product and construction, this is hardly controversial. But in contrast, it is not at all obvious to the new net-citizens that a comparatively small group of software-producers should earn large sums of money by selling expensive digital information which is incredibly easy and cheap to copy and distribute.

The special conditions that pertain within the informational economy mean that the production costs, storage and distribution of digital products are practically negligible, a fact which would benefit the whole of the net community if the right of ownership of information was dismantled. An unavoidable analogy is how the workers in the factories of early capitalism accepted the feudal right to ownership of land as being natural, but gradually began to question the fact that capitalism's means of production, the factories and their inventories, should belong to the bourgeoisie. We are all too aware of what violent and bloody conflicts this struggle for the means of production led to. There is no reason whatever to believe that the informational class-struggle will be any calmer or more peaceful.

The problem for anyone protecting restrictive copyright-legislation is that the increasingly dominant mobilistic ways of thinking are negative to the very idea that a certain combination of ones and zeros could belong to any particular person or organisation, which means that this ownership will in time come to seem more and more "unnatural". All this legislation and all attempts to implement it will be regarded as the protection of illegitimate special interests, which will in turn lead to the fragile alliance between netocratic entrepreneurs and capitalist investors coming under severe pressure. Those protecting the right to immaterial ownership will be forced to fall back on a collapsing political structure: the nation state, which will have no economic or even practical capacity to call out the police whenever immaterial rights of ownership are transgressed.

How could any police force or group of politicians set about closing web-sites that operate from some isolated island off Africa or in the Pacific, or even from a nation without control of its virtual domains (the Soviet Union's domain name .su was, for instance, quickly snapped up by opportunistic hackers after the collapse of the Soviet Union)? Besides, there is every reason to question what moral right the informational economy, which is increasingly avoiding national taxation, might have to demand any protection from the state. As a result of this development, netocratic entrepreneurs will be forced to construct new systems to protect their desirable information. They will encrypt the ones and zeros before distribution; they will build increasingly sophisticated firewalls, virtual walls, around their activities to protect against eventual break-ins. The netocrats will educate their own guards and create their own networks carrying constantly updated information about electronic pirates and traffickers in stolen goods. In this way they will make themselves independent of the state and will accelerate, ironically enough, their own acquisition of power.

The conflicts of the informationalist era will therefore not take place between nation states fighting over tracts of land of questionable value. Instead we can see how ideological and economic conflict between different netocratic groups, in more or less loose alliances with consumtarian rebel movements, is developing. The dividing-line runs between the netocracy which is protecting what it regards as its rightful ownership of the information that forms the basis of the group's power and status, and netocratic class-traitors who regard every form of hindrance to the spread of information as immoral, and instead see the maximal expansion of the organisational non-zero-sum game as the core-value of the new age. One clear example of this sort of internal netocratic conflict was the frenetic struggle to be the first to present a complete map of the human genome. The two sides consisted of an international consortium of academic research institutes, HGP, which claimed to be working for the general good and without profit-motivation, and a purely commercial company, Celera, whose business plan is to restrict and make money from specific patents in the genetic arena.

But because the central value of the informational economy does not lie in information itself, but in the sorting and combination of information, the most powerful netocrats need not concern themselves with ownership of copyrights and patents. Nor do they need to invest time and effort in the construction of encryption programmes or firewalls. The ability to network and get an overview of large amounts of information which is sought after by everyone cannot be copied or stolen; the owner is threatened by nothing but the possibility that someone else will prove themselves more talented. And this will form the basis for the growth of an alternative netocracy, an elite which will base its power and status upon entirely different factors than copyright and patents (and therefore ownership of the means of production).

This new group, the eternalists, will sympathise and collaborate with the industrious members of the same class, the nexialists, only when it is in the interests of the group itself. In other cases, when opposition splits the netocracy, it might well betray its own class and make common cause with the consumtariat, rather like academics with leftwing sympathies in late-capitalist society at least on occasion tried to make common cause with the working class against the class in power, to which they themselves de facto belonged. The academic left will have a successor in the eternalistic netocracy which regards certain of the ruling elite's actions as immoral and offensive. This is an attitude it can afford to adopt even when the situation in question does nothing to enhance its own interests.

All of this, together with the new view of the relationship between production and consumption, gives a clearer picture of how the class-conflicts of informational society will manifest themselves. The most fundamental form of consumtarian resistance will be to refuse to produce desire, by boycotting both adverts and technology, and by withdrawing from the informational economy as far as possible. The activist form of resistance will be to attack the key functions of the netocratic entrepreneurs, the nexialists: copyright and patents. Revolutionary resistance will find expression in both virtual and physical violence. Every form of protective wall around highly valued information will become a target.

What the struggle is ultimately about is no longer control of production, but control of consumption. One characteristic of the netocrat is that he controls his own desires and exerts strong influence on others', while the consumtariat's production of desire is directed from above. The consumtariat will become resistant to the power of the netocracy at the moment it no longer accepts this state of things. The cheerleaders of "the new economy" believe, or pretend to believe, in the emergence of a collective, joyous realm where social tensions will be dispelled by the winds of change and everything will be resolved to the good thanks to large amounts of information. This is far too naïve an attitude, probably often a simulated one. On the contrary, tensions will increase in the electronic class society which is developing under informationalism. History is not dead, it is being resurrected in the present in an entirely new guise. And in the same way as this resurrection of history demands new players, any credible observation of informational society demands new observers.

Futurica Glossary

Concepts common in the philosophical works
of Alexander Bard & Jan Söderqvist.

ABSOLUTE MOBILISM: A natural romantic utopianism, a dream of pure thought superior to all culturally conditioned categories, free-flow philosophy. Unfortunately impossible to practice and communicate, because the medium, language, requires an eternalisation of thought.

ATTENTIONALISM: The value system which is gradually replacing economism as the informationalist paradigm is becoming established. Attention is determined by the interplay between the awareness and the credibility a person or network succeeds in creating.

BIFURCATION: In geometry, the point where a line divides or branches into two lines. Launched in philosophy by Gilles Deleuze, since when it has been used in the wider context for conditions in which conhesive eternalisations burst, are mobilised and have to be eternalised once more in order to be made comprehensible.

CAPITALISM: The social paradigm which was introduced with the printing press and made its broad breakthrough thanks to effective one-way communication with a mass audience created by these new media. The concept is therefore much more than what is sometimes referred to as capitalism in everyday speech, in other words: the market economy, economism, et cetera.

CONSUMTARIAT: The underclass of information society, a proletariat of consumption which lacks the power and initiative to produce its own social identity, and which is therefore condemned passively to consume a pre-produced identity.

DEMONOLOGY: Society's production of enemies and threats in order to reinforce internal cohesion and define the collective identity.

DIVIDUAL: A human being perceived as divisible rather than indivisible (the individual). The dividual nurtures a multiplicity of identities, regarding none of these as more 'real' or 'original' than any other, and allows the different sides to dominate according to context, whereas the individual strives for an integrated personality.

ÉTATISM: The supreme political ideology of the capitalist paradigm, of which conservatism, liberalism and socialism are closely-related varieties. The whole of political thought stems from the idea of the nation state as the principle social institution and self-evident centre of power.

ETERNALISATION: A freezing of a chosen piece of absolute movement in the chaos of the world, necessary to make life comprehensible.

ETERNALISM: The philosophy which through eternalisations and meta-eternalisations produces stability in full awareness of the fallacy of stability, and in close interaction with the entirely different but nonetheless unavoidable mobilism.

ETHICS: Original meaning 'habits' in Greek. Since Baruch Spinoza's definition in the 1600s, used philosophically as synonymous with presumed subjective values based on a monistic and secularised worldview. Operative from the oppositional pair constructivity v. destructivity.

THE EVENT: The glittering goal of the networking collective. The Event arises when productivity in the network reaches maximal levels and lasts as long as this state can be maintained. It replaces the capitalist era's idea of Progress as the subject's desired reward for hard work.

EVENTOLOGY: The theory of the conditions and social functions of the Event.

FEUDALISM: The social paradigm linked to the information-technological revolution of written language c.3,000 BC, and which stretched until the next revolution, that of the printing press. Feudalism includes the feudal society of the European Middle Ages, but is also a wider concept.

FICTIVES: The instruments that thinking itself produces in the dialectic between eternalism and mobilism, when the first set of eternalisations are mobilised and thereby set in motion, meaning that they have to be eternalised again.

THE GLOBAL EMPIRE: The idea of a global order – political, cultural and economic. A vision which appears in many different forms and with many different focuses in the history of philosophy, all of them supporting the view that the global platform for the exercising of power is unequivocally superior to regional interests.

GLOBALISATION: A genuine globalisation comprises infinitely more than merely free trade: a truly global, cultural and political arena for common problem-solving and mutual enrichment in all possible areas.

GLOBALISM: The political ideology which follows eternalism, and the activist movement towards a global order. Identity: Humankind's primary social needs, more or less satisfied by the collective subject's production of identity.

INFORMATIONALISM: The social paradigm which is replacing

capitalism in conjunction with the information-technological revolution of interactive media. Synonyms: Information society and Network society.

INTERACTIVITY: Two-way communication, media technologies for mutual information exchange where the roles of producer and consumer are constantly being swapped.

MATHEMES: Philosophical diagrams used to illustrate relationships between phenomena, for instance the relationship between credibility and awareness in the production of attention. Frequently used within Lacanian psychoanalysis to illustrate how consciousness regards itself in order to function.

MEMETICS: The theory of memes and their fight to survive and reproduce. Memes are the cultural equivalent of genes.

THE METAPHYSICS OF PARADOXISM: The idea of the never-ending dialectic between eternalism and mobilism as the single but entirely sufficient metaphysical basis for both thought and life itself.

MOBILISATION: The return of single eternalisations to a state of movement and new, meaning-creating relations to one another.

MOBILISM: The philosophy which sees through the fiction of eternalisation (and totalism's ambitions of totality), and returns thought from artificial order to the productive chaos of the next level, all the time in close interaction with the entirely different but nonetheless unavoidable eternalism.

MORALS: Original meaning 'habits' in Latin. Since Baruch Spinoza's definition in the 1600s, used philosophically as synonymous with presumed objective values based on a dualistic and theological worldview. Operative from the oppositional pair good v. evil.

NIHILISM: The understanding that life lacks both divine and natural meaning, as a result of which the search for meaning has always been and will always be a production of meaning. Eternalism differentiates between naive, cynical and affirmative nihilism.

THE NET: The metaphysical aspect of networking as an idea, and therefore something much broader and more abstract than any specific network or any specific technology.

NETOCRACY: The dominant class of informationalism, in other words: the elite which attains power and status through its unique capacity to create attentional value, and which produces and controls its own social identity, primarily through intensive networking.

NETWORK DYNAMICS: The branch of socioanalysis which studies how relations, hierarchies, transparencies, topographies, non-zero-sum games, zero-sum games and minus-sum games arise and function within different networks.

NON-ZERO-SUM GAME: A concept taken from game theory.
A non-zero-sum game is characterised by mutual benefit from the production of increased value: every player has good reason to hope for advantage. The result is greater than the sum of the parts. In a zero-sum game, you can only win what someone else loses. The result is exactly the same as the sum of the parts. In a minus-sum game, the productivity is negative, the result is less than the sum of the parts, as a result of which all players risk net losses.

PARADOXISM: The dialectic aspect of eternalism.
The Paradox of Metaphysics: The never-ending and apparently insoluble conflict between eternalism (stability philosophy) and

mobilism (movement philosophy). While the former stems from being, the latter is based upon becoming. The solution to this conflict suggested by Bard & Söderqvist proposes a metaphysics of paradoxism, in which the relationship between eternalism and mobilism forms a circular but simultaneously open dialectic which itself constitutes the metaphysical foundation.

PLURARCHY: A political state where majority decisions have lost all weight, as a result of which all subordination is based upon some form of voluntary submission.

REPLICATORS: Molecules or cultural concepts (genes or memes) with the capacity to produce exact copies of themselves. Occasionally these copies are not exact, which gives rise to mutations, which in turn facilitate evolution and natural selection.

SCHIZOANALYSIS: Psychoanalysis modified for the needs of the dividual, intended to liberate a large but still manageable multiplicity of personalities, and to counteract every tendency towards restrictive uniformity.

SOCIAL TECHNOLOGIES: Principles for social interaction, consciously or unconsciously applied as instruments with which to steer the collective in a certain direction.

SOCIOANALYSIS: Schizoanalysis transferred to the level of the collective subject. Informational society's politics are applied socioanalysis, as a result of which the voters of the capitalist era will be replaced by socioanalysis, whose relationship to politicians/analysts will be characterised by interactivity and the plurarchic form of voluntary subordination.

SOCIOMETRY: The endlessly complex measuring and comparison of social status on attentional terms. Constructed upon sociograms, diagrams of who knows and interacts with whom and in which ways in informational society.

TOTALISM: The tradition of thought which has dominated western philosophy since Plato's day, and which aims to develop a form of thought which encompasses all of existence as it actually is, which, from a mobilistic perspective, is absurd, given that the world of existence is by definition different to the conceptual world.

TRANSRATIONALISM: A modified and pragmatic rationalism which accepts and stems from the in-built limitations of rational thought. Transrationalism is interested in what is functional rather than in objective truths and possible correspondences between thought and existence.

VIRTUAL NOMADIC TRIBES: Netocratic collectives in the wider sense. They are virtual because they are primarily organised via interactive media technologies, nomadic because mobility is a virtue and geographical location uninteresting, and tribes because they tend to form according to primitivism's tribal structures.

THE WORLD STATE: The Global Empire translated into judicial and political practice, meaning global jurisdiction and a global monopoly of violence, which in turn means the final death-throes of the sovereignty of the nation state.